Management Technologies for Metal Mining Influenced Water

Basics of Metal Mining Influenced Water

Volume 1

Edited by Virginia T. McLemore

Published by

Society for Mining,
Metallurgy, and Exploration, Inc.

Society for Mining, Metallurgy, and Exploration, Inc. (SME)
8307 Shaffer Parkway
Littleton, Colorado, USA 80127
(303) 973-9550 / (800) 763-3132
www.smenet.org

SME advances the worldwide mining and minerals community through information exchange and professional development. SME is the world's largest association of mining and minerals professionals.

ISBN-13: 978-0-87335-259-8

On the Cover: A 1993 photo showing acid drainage from the Terrero waste rock pile draining into a wetlands before entering the Pecos River in New Mexico.

Library of Congress Cataloging-in-Publication Data

Basics of metal mining influenced water / Edited by Virginia T. McLemore.
p. cm. -- (Management technologies for metal mining influenced water ; v. 1)
Includes bibliographical references and index.
ISBN 978-0-87335-259-8
1. Mine drainage. 2. Water--Pollution. I. McLemore, Virginia T. II. Society for Mining, Metallurgy, and Exploration (U.S.)
TD899.M5B37 2008
628.1'6832--dc22

2007051835

Contents

Editorial Board and Contributors

Senior Editor

Virginia T. McLemore, New Mexico Bureau of Geology and Mineral Resources, Socorro, New Mexico

Associate Editors

Charles Bucknam, Newmont Metallurgical Services, Englewood, Colorado
James J. Gusek, Golder Associates, Denver, Colorado
Harry Posey, Shell, Denver, Colorado

Contributors

David Bird, State of Colorado, Denver, Colorado
Devin Castendyk, State University of New York, Oneida, New York
Terry Chatwin, Utah Engineering Experiment Station, Salt Lake City, Utah
Mark Cowan, U.S. Army Corps of Engineers, Sacramento, California
Ted Eary Montgomery Watson, Fort Collins, Colorado
Linda Figueroa, Colorado School of Mines, Golden, Colorado
Jim Jonas, Virotec, Westminster, Colorado
Mike Klosterman, U.S. Army Corps of Engineers, Washington, D.C.
Kim Lapakko, Minnesota Department of Natural Resources. St. Paul, Minnesota
Glenn Miller, University of Nevada, Reno, Nevada
Andrew Nicholson, Geomega, Boulder, Colorado
Brian Park, MSE, Butte, Montana
Doug Peters, Peters Geosciences, Golden, Colorado
Mark Peterson, Montana Tech, Butte, Montana
Nick Reiger, U.S. Bureau of Land Management, Washington, D.C.
Carol C. Russell, U.S. Environmental Protection Agency, Denver, Colorado
Kathy Smith, U.S. Geological Survey, Denver, Colorado
Ed Trujillo, University of Utah, Salt Lake City, Utah
Dennis Turner, Arizona Department of Environmental Quality, Phoenix, Arizona
Dirk J.A. Van Zyl, University of British Columbia, Vancouver, British Columbia
Rens Verburg, Golder Associates, Redland, Washington
Bruce M. Walker, Chevron, Questa, New Mexico
Bill White, retired, Heber, Utah
Dave Williams, U.S. Bureau of Land Management, Butte, Montana

Additional Contributors

Lynn Brandvold, retired, Socorro, New Mexico
Ibrahim Gundiler, New Mexico Bureau of Geology and Mineral Resources, Socorro, New Mexico
Larry Todd, Freeport McMoran, Safford, Arizona
Tom Wildeman, Colorado School of Mines, Golden, Colorado

In addition, figures and photographs were provided by Virginia T. McLemore, Meghan Jackson, Leo Galbadon, and Tom Kaus (of the New Mexico Bureau of Geology and Mineral Resources). The U.S. Office of Surface Mining Reclamation and Enforcement, the New Mexico Bureau of Geology and Mineral Resources, and the International Network for Acid Prevention provided direct funding; and the participants' companies, institutions, and agencies provided in-kind funding.

Preface

Basics of Metal Mining Influenced Water serves as an introduction to a series of six handbooks on technologies for management of metal mine and metallurgical process drainage. The other five handbooks are *Mitigation of Metal Mining Influenced Water; Mine Pit Lakes: Characteristics, Predictive Modeling, and Sustainability; Geochemical Modeling for Mine Site Characterization and Remediation; Techniques for Predicting Metal Mining Influenced Water;* and *Sampling and Monitoring for the Mine Life Cycle.*

These handbooks are a volunteer project of the Acid Drainage Technology Initiative–Metal Mining Sector (ADTI-MMS). The work was directed by the ADTI-MMS Steering Committee, a technically focused consensus group of volunteer representatives from state and federal government, academia, the mining industry, consulting firms, and other interested parties who are involved in the environmentally sound management of metal mine wastes and drainage quality. The mission of ADTI is to identify, evaluate, develop, and disseminate information about cost-effective and environmentally sound methods and technologies to manage mine wastes and related metallurgical materials for abandoned, inactive, active, and future mining and associated operations, and to promote understanding of these technologies.

These handbooks describe the technical aspects of sampling, monitoring, mitigation, and prediction programs of the mine life cycle. The audience for these technical handbooks includes planners, regulators, consultants, land managers, researchers, students, stakeholders, and anyone with an interest in mining influenced water (MIW).

Although numerous handbooks, both technical and nontechnical, are available about acid drainage and the technologies used to sample, monitor, predict, mitigate, and control acid drainage and other mine wastes, most of these handbooks relate primarily to acid drainage from coal mines. But not all adverse drainage from metal mines is acidic; some neutral pH waters can be detrimental to the environment. In the introduction (Chapter 1), we explain that the use of the term *mining influenced water* refers to all waters affected by mining and metallurgical processing, which includes wastes from historic mining operations. This term resolves much of the confusion that exists from using *acid mine drainage* for cases in which drainage comes from mines but is not acidic. The ADTI-MMS handbooks address all MIW, not just acid drainage.

Unlike many other handbooks, this ADTI-MMS introductory handbook begins by reminding us of the importance of mining (Chapter 2). Mining is essential to maintain economic wealth because it is one of the ways that new wealth is created (fishing, farming, and forestry are others). Without a continued supply of natural resources to manufacture and produce products that society demands, society cannot continue to improve and grow technologically. Production of all minerals, including coal, metals, industrial minerals, and other mineral commodities, has increased dramatically over the last century and will continue to increase in the future. Mining companies realize that they must meet the mineral needs of our society while protecting the health, safety, and welfare of its workers, the surrounding community, and the environment (the subject of these handbooks).

In addition, the specific functions of metal and coal mining result in different types of mining and processing methods, which result in different types of MIW that have to be sampled, monitored, modeled, and mitigated (described in Chapter 3). The ADTI-MMS handbooks focus on reviewing the reclamation and prevention technologies and philosophy of mining within and beyond the requirements of government regulations.

These handbooks discuss another important aspect of mining today compared with mines of the past. Planning a new mine in today's regulatory (and often political) environment requires a different philosophy in terms of designing a new or existing mine or expanding operations for ultimate closure. The holistic philosophy presented in the ADTI-MMS handbook series maintains that sampling, monitoring, prediction, and mitigation programs should be designed to take into account all aspects of the mine life cycle (described in detail in Chapter 4). This philosophy is new, as the miners and regulators in the past rarely planned for mine closure. Because complying with environmental regulations is the job of everyone at all stages of mining, closure planning should ideally begin during the exploration and feasibility phases and be augmented throughout development, mining, milling, and, finally, closure and postclosure.

This book includes a glossary compiled by Doug Peters, with the help of Jim Gusek, Carol Russell, and Kathy Smith; and chemical equations, most of which are in the appendix instead of the text for easier reading and understanding. This introduction handbook benefited from the expert review and multiple revisions by the associate editors, Harry Posey and Jim Gusek. Finally, without the leadership, support, encouragement, and tenacity of associate editor Charles Bucknam, this handbook series would still be in a draft version instead of this publication series.

CHAPTER 1

Introduction

In shaping the course of civilization, mining plays a unique and important role in the world economy and the future of society by filling the persistent demands for raw earthen materials and native metals. Because mineral needs are present in all societies, mining contributes essential products for their sustained economic future. Metals and industrial minerals are used in every sector of construction and manufacturing. While energy minerals provide electricity and fuels for all aspects of industry and society, minerals for fertilizers and pesticides are necessary for agriculture. Indeed, mining is the foundation of civilization, no matter how primitive or advanced.

The production and flow of minerals in the United States and the world has increased dramatically in the last century (Figure 1.1; Wagner 2002). Not only is the United States a major producer of aggregate, iron, copper, lead, zinc, gold, silver, molybdenum, and industrial minerals, but it also imports these and other commodities (S.S. Smith 2001; Wagner 2002). As population increases worldwide and as people demand an increasingly better quality of life, production and consumption of minerals will increase in the future.

The value of mining multiplies in the economy in a number of ways. Mines provide benefits to the communities where they are located by contributing wages (mine staff are highly trained and educated, and their wages average higher than other industrial and social sectors); by generating economic activity due to purchases of goods and services; and by supplying taxes, royalties, and fees to local, state, and national governments.

Environmentally responsible mining and mineral processing are important to national economies and to quality of life as indicated by the following quote from the International Institute for Environment and Development: "One of the greatest challenges facing the world today is integrating economic activity with environmental integrity and social concern. The fulfillment of 'needs' is central to the definition of sustainable development" (IIED 2002). A large part of the contribution of mining to sustainable development is the continuing flow of minerals by mining while maintaining the well-being of the physical and social environment as much as possible.

One important aspect of mine planning in a modern regulatory setting is the philosophy—or the requirement in most cases—that new mines and mine expansions must have plans and designs for closure. This philosophy is relatively new, especially in the metal mining industry. Mine development in the distant and even not-too-distant past commonly did not consider mine closure, except perhaps to plan for safeguards and contingencies. Today, mine closure planning is necessary not only for safety but also for environmental reasons. Economics also plays a role, because adverse environmental impacts can cause significant and serious economic fallout.

Ideally, closure planning begins during exploration and feasibility phases of mining, when environmental geochemistry and predictive tests can provide early signals of future water problems. Some plans need to be adjusted or augmented throughout the development and excavation phases of mining, when rocks are maximally exposed and the best and most representative samples can be collected. In the absence of careful closure planning, environmental problems left

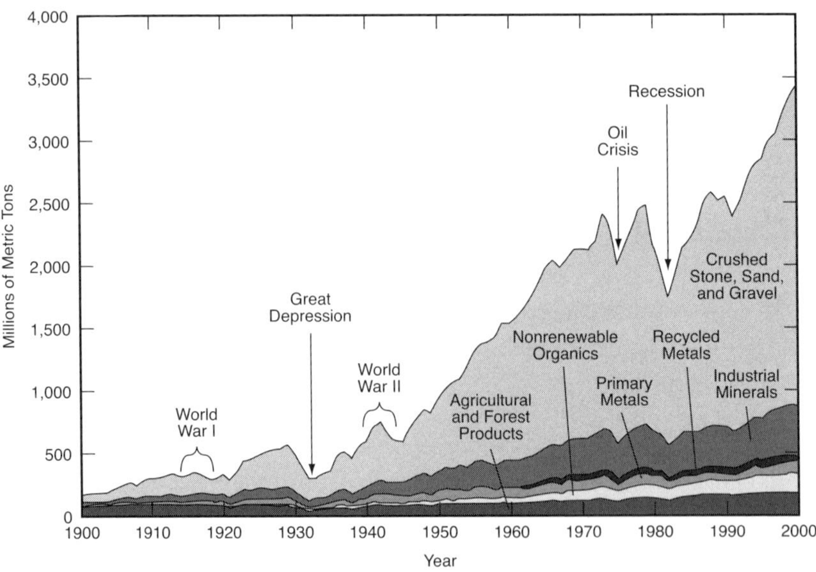

Source: Wagner 2002.

FIGURE 1.1 U.S. flow of raw materials by weight, 1900–2000

behind from mining operations can result in unintended socioeconomic costs, presenting problems for the mine's reputation and unwelcome burdens on local resources and society.

One of the major problems to address in protecting the physical environment is acid drainage. The chemistry of drainage from materials generated by mining and metallurgical processing is controlled predominantly by local mineralogy and climate and is affected (sometimes significantly) by environmental features such as hydrology, slope, and aspect. More importantly (in the context of this series of six handbooks by the Acid Drainage Technology Initiative–Metal Mining Sector [ADTI-MMS]), the chemistry of drainage from mining and metallurgical materials can be usefully affected by human-made features such as surface water runoff controls, waste rock placement, liners, covers, and water treatment, to name a few. The chemistry of the material identifies elements available for release to drainage. Mineralogy and climate, particularly precipitation rate and temperature, influence weathering rates; and climate, as well as location in the watershed, also influences dilution rates. Mining conditions affect the amount of contact (and therefore the interaction) between rocks and water. As these handbooks emphasize, deliberate choices made prior to and during mining, such as where and how to place waste rock and whether to install liners or caps, can greatly affect drainage water quality, and the effects of such choices are most beneficial when made early in the mining life cycle.

Because rock chemistry and climate vary among mines, plus a need to resolve mine water problems locally, researchers have developed a range of techniques for resolving mining influenced water problems. These attempts to understand, model, predict, monitor, control, and treat acid- and metal-bearing drainage in waters affected by mining or metallurgical activities are the foundation of the ADTI-MMS series, Management Technologies for Metal Mining Influenced Water.

Collectively, waters affected by mining and metallurgical processing are referred to as mining influenced water (MIW), which includes waste piles from historic mining operations. Although some jurisdictions use the term *MIW* only for waters that are adversely affected, these handbooks use the term in the more general context. The term resolves much of the confusion that exists from using "acid mine drainage" for cases where drainage derives from mines but is not acidic. More specific terms such as *acid rock drainage, acid drainage, alkaline drainage,* and *high metals drainage* are dealt with on a case-by-case basis. MIW is reserved for waters from mines and metallurgical processes that have been affected, adversely or not, by mining and processing.

No single method is available to resolve MIW problems, because conditions vary from site to site, including water chemistry; local climatic, geological, and topographic conditions; mining methods; and other extrinsic factors. Selection of techniques for solving mine water problems requires evaluation of these variables and a site-specific assessment that considers factors such as cost, potential engineering effectiveness, the problem's magnitude, and maintenance needs.

The volume of literature on MIW drainage is great, and several authoritative guides present a wealth of water treatment techniques for MIW problems. The Acid Drainage Technology Initiative–Coal Mining Sector (ADTI-CMS) has compiled two handbooks on acid mine drainage as it relates to coal mining (Skousen et al. 1998; Kleinmann 2000), and there are scores of books and guidance documents in the literature, including from the Mine Environment Neutral Drainage, Mitigation of the Environmental Impact from Mining Waste, the Australian Centre for Minerals Extension and Research, and many company and agency documents. To further the science of MIW treatment and control and to help disseminate technical information, the ADTI-MMS is developing this series of six handbooks on MIW relating to metal mining, metallurgical processing, and selected industrial minerals mining in the United States.

The ADTI-MMS handbooks on MIW are technical documents, intended to present an overview of the best documented science and methodologies for sampling, monitoring, prediction, mitigation, and modeling of drainage from metal mines, pit lakes, and related metallurgical materials based on current scientific and engineering practices. This series of handbooks is not intended to foster or substitute for regulatory requirements, but rather to provide a review of tools for solving technical problems. They address most types of MIW control and treatment and include general design and performance criteria, as well as historic examples and case studies. The handbooks provide design details associated with successful mining operations and case histories of several mine failures, evaluate cost effectiveness of various options, and identify research needs. The series will aid users in selecting technologically effective and economical sampling, monitoring, prediction, and mitigation protocols, and provide useful information on prediction and modeling for particular geological and environmental situations. The handbooks are meant for anyone with an interest in mining or its associated environmental operations and can be useful for planners, regulators, consultants, land managers, researchers, students, and stakeholders.

This introductory handbook, *Basics of Metal Mining Influenced Water*, focuses on aspects of metal mining and metallurgical processing needed to provide a background for the rest of the handbooks. Basic information is provided about some of the major physical and chemical relationships between mining, climate, environment, and mine waste drainage quality; a description of metal mining, including how metal mining differs from coal mining; and the types and phases of metal mining. In ensuing chapters, the environmental impacts from metal mining and metallurgical materials are reviewed and the hydrologic cycle, the importance of climate, and the geochemistry of MIW are described. Other related issues, specifically erosion and sedimentation, stability issues, dust emissions, and other chemicals of concern, are also discussed. While a few formulas are included in the text, others are in the appendix.

The second volume in the series, Mitigation of Metal Mining Influenced Water, surveys methods for avoiding or controlling adverse drainage and mitigation of MIW. It discusses the physical environments that affect mining and the social attitudes and regulatory conditions that influence mining, mine plans, and mine closure. The handbook presents specific design and operational strategies that address avoidance, control, and treatment.

The third handbook, *Mine Pit Lakes: Characteristics, Predictive Modeling, and Sustainability,* summarizes some of the current knowledge about mine pit lakes, locally a large-scale problem at some mine sites. Physical and chemical features of pits, host rocks, and waters are discussed along with sampling approaches, modeling techniques and predictions, remediation approaches currently being practiced, and considerations for postclosure management as a resource. Wherever surface mines draw down local water tables and the local rocks are sulfide bearing, there is a relatively high likelihood that water in the resulting pit lake and surrounding groundwater will be acidic and problematic. This handbook gives basic information on how to understand and predict pit lake chemistry, how to effectively manage a pit lake, and under what circumstances it can be reasonable to backfill either above or below the resultant water table.

The fourth volume, *Geochemical Modeling for Mine Site Characterization and Remediation,* reviews geochemical modeling applied to MIW. Mine waste management decisions are sometimes based on predictions of mine drainage quality. Although laboratory-predictive testing results are supplied by mining companies and reviewed by permitting and land-management agencies to help forecast mine waste drainage quality, there is no consensus on how well these tests can predict future drainage quality on a site-specific basis. Mathematical models of the chemical, biological, or physical processes involved in mine wastes are tools that can be used to understand and estimate the behavior of these systems into the future. Two categories of geochemical models are generally available: (1) those that describe geochemical and transport (mass, flow, and energy) aspects of the system, and (2) those that describe only the geochemical aspects. That handbook reviews the common models and provides guidelines to help the nonspecialist evaluate these models and their applications.

The fifth handbook, *Techniques for Predicting Metal Mining Influenced Water,* reviews laboratory test methods for predicting drainage quality and for guiding technology development to manage drainage quality from mining and related metallurgical materials for future, active, and abandoned mines and associated metallurgical operations.

In the sixth, and final, handbook, *Sampling and Monitoring for the Mine Life Cycle,* holistic sampling and monitoring are described, expressing the philosophy that sampling and monitoring of any single feature are the most useful parts of an integrated plan that includes the entire life cycle of a mine from exploration through closure. That handbook discusses procedures and techniques to sample and monitor environmental impacts associated with mine waste drainage. Although numerous sampling and monitoring techniques have been developed for mine sites throughout the world, there is not much literature describing holistic philosophies for sampling and monitoring. The handbook presents some of the commonly used techniques, including ones likely to provide useful data, and their cost-effectiveness.

BACKGROUND OF ADTI

The ADTI was created in 1995 by the National Mine Land Reclamation Center, National Mining Association, Office of Surface Mining Reclamation and Enforcement, U.S. Environmental Protection Agency (EPA), Bureau of Land Management, U.S. Geological Survey, and Interstate Mining Compact Commission. Its original mission was to identify, evaluate, and develop cost-effective and practical acid mine drainage technologies and to address other drainage-quality issues related to mining (Hornberger et al. 2000). The focus of ADTI has subsequently evolved to identify, evaluate, develop, and disseminate information, and promote understanding of cost-effective and environmentally sound scientific methods and technologies to manage MIW and related processing materials.

In 1999, ADTI was expanded through the addition of the Metal Mining Sector, after it was decided that a separate noncoal sector was needed because of the different geology and drainage chemistry associated with coal and noncoal mines. ADTI now addresses MIW at metal mines and related metallurgical operations, including pit lakes, through ADTI-MMS, and addresses MIW from coal mines through ADTI-CMS. These groups address historic (or abandoned), active, and future mines.

ADTI-MMS has assembled leading, national experts in acid drainage, metal mining, metallurgical processing, geology, geochemistry, mineralogy, analytical technology, and drainage quality control technology to identify up-to-date prevention and remediation technologies and to identify promising technologies for focused research. ADTI is a coalition of state and federal agencies, industry, academia, and consulting firms.

DEFINITIONS

In these handbooks, common scientific definitions are used wherever possible. Specific legal definitions of scientific terms, regardless of legal entity, are not employed unless noted. Terms such as *soil*, which have slightly different definitions in several science and engineering disciplines, are described as applied. Where common dictionary definitions suffice, terms are not defined. A glossary of terms used throughout the handbook series can be found at the back of this book.

A *mineral occurrence* is any locality where a useful mineral or material is found. A *mineral prospect* is any occurrence that has been developed by underground or aboveground techniques, or by subsurface drilling to determine the extent of mineralization. These two terms do not have any resource or economic implications. A *mineral deposit* is any occurrence of a valuable commodity or mineral that is of sufficient size and grade (concentration) for potential economic development under past, present, or future favorable conditions. An *ore deposit* is a well-defined mineral deposit that has been tested and found to be of sufficient size, grade, and accessibility to be extracted (i.e., mined) and processed for profit at a specific time. Thus, the evaluation of the size and grade of an ore deposit changes with time as the economic conditions change.

Mineral deposits and especially ore deposits are relatively rare and depend on certain natural geologic conditions to form. The requirement that ore deposits must be extracted at a profit makes them much rarer than mineral deposits. Since an ore deposit is a subset of a mineral deposit, the term *mineral deposit* will be used in most instances in these handbooks.

The formation of mineral occurrences and deposits require a source of elements (usually aqueous fluids), a transport mechanism (for instance, gravity-induced fluid flow), a concentration mechanism (to increase or concentrate the valuable minerals into an economic deposit), and preservation of the deposit (Figure 1.2). Subsequent geologic processes, such as supergene enrichment and erosion, can either modify or destroy mineral deposits. Mineral deposits can be divided into two basic types: metals and industrial minerals. These handbooks address MIW issues associated with metallic and selected industrial minerals deposits.

A *mine* is any human-made opening or excavation in the ground created for the purpose of extracting minerals. In a general sense, a mine includes any mineral-related opening or excavation in the ground, even if no minerals were extracted. Some restrict the definition of a mine to include only those excavations where a commodity was actually produced. This restricted definition is more difficult to use in practice, because production records are not available for most historic

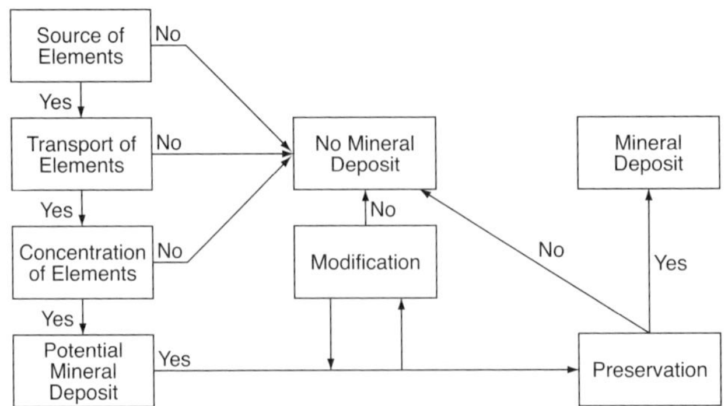

Source: Adapted from Tilsley 1990.

FIGURE 1.2 Conceptual model for the formation of a mineral deposit

FIGURE 1.3 The Argo Tunnel near Idaho Springs, Colorado, typically produces water flow rates ranging from 200 to 300 gpm, depending on the season of the year. The pH ranges between 2.7 and 3.2, and the water contains high concentrations of zinc, iron, sulfate, cadmium, copper, and manganese.

mines. Some mines may have produced only a few pounds of a commodity, whereas other mines were used only for exploration. However, whether or not a mine actually produced commodities is not as important as the environmental consequences of that particular opening or excavation.

Acid drainage (AD) results from the oxidation of sulfide minerals in rocks exposed to air and water. It can be either a natural occurrence or a consequence of mining-related activities, such as drainage tunnels (Figure 1.3). When the drainage source is a mine, AD is also referred to as acid mine drainage. The term *MIW* is considered more inclusive of all types of mine drainage, because metals and other contaminants can be transported by neutral and alkaline waters from materials originating at a mine. Water-draining mine tunnels were once a common feature of mining, especially in areas of steep topography. In addition to lowering the local water table, draining tunnels were intended to reduce flooding or provide access to underground mines from beneath. By lowering the water table, draining tunnels replaced underground water pumps and thereby lowered the cost of mining. Unfortunately, by lowering the water table, many draining tunnels provided oxygen and water to previously unexposed sulfides, thus generating acidic waters.

CHAPTER 2

Characteristics of Mining and MIW

Mining is the economic extraction of raw materials from the earth, and, as a major industry, creates wealth by providing the materials needed to sustain our way of life. Typically it is divided into three sectors: coal mining, industrial minerals, and metal mining. Industrial minerals are any rock, mineral, or other naturally occurring material of economic value, excluding metals and energy minerals. Typically, they do not create acid drainage or other adverse water quality issues, and, if they do, those issues are covered in this series of handbooks. Because ADTI-CMS has written handbooks on coal mining, this series discusses metal mining, although a comparison of coal and metal mining illustrates how their different geology, mineralogy, and processing techniques affect mine and processing water chemistry.

HOW METAL MINING DIFFERS FROM COAL MINING

Many technologies have been developed to remediate acid drainage from coal mines. However, many of these technologies do not always apply to metal mines because of the many differences between coal and metal mining. Acid drainage is a characteristic of many (fortunately not all) coal mines, because of the common presence of iron sulfide minerals in coal that produce acidity upon exposure to weathering. Coal mines in the eastern and central United States, for instance, commonly have drainage with high acidity and low pH, and high concentrations of iron, sulfate, manganese, and aluminum. In some cases, other problematic trace elements are present, such as arsenic and selenium. Western coal mines, however, mine low-sulfur coals and do not generally have significant drainage quality problems.

Some metal mine drainages throughout the world have acid, sulfate, and metal signatures that are similar to eastern U.S. coal mine drainage. However, metal mine drainages (at least the problematic ones) also tend to have additional problematic metals, because the oxide and sulfide minerals in metal mining ores, waste rocks, and metallurgical processing products are more complex than coal. Zinc can be the most common problematic metal, but arsenic, cadmium, copper, lead, and/or mercury are frequently present; other metals, such as antimony, cobalt, nickel and/or selenium, occur more rarely.

Acid formed in mine drainages can be passively neutralized by contact with carbonate minerals such as calcite and dolomite (or their rock equivalents, limestone and dolostone). These carbonate minerals can also prevent the formation of bacterially accelerated acid drainage through maintaining an elevated pH. Circumneutral pH, however, will not necessarily ensure sequestration, as metal solids of dissolved metals may occur as oxyanions, so both neutralized and naturally alkaline metal drainage can be problematic. Such drainage can derive from both mined metal and coal deposits, especially those hosted or in close contact with carbonate strata. These neutral and alkaline drainages, which can cause difficulties because of dissolved metals, but not acid content, seem to be more commonly associated with metal mine and metallurgical processing wastes than those from coal.

When considering water quality mitigation, it can be useful to consider how local geology and mineralogy have affected the generation and, perhaps, neutralization of acid and how those same features have affected metal content of the drainage. Partly for that reason, ADTI is divided into coal and metal mining sectors, and the two sectors generally distinguish the drainage chemistry of coal deposits from that of metal deposits.

At aboveground metal mines, the majority of the mined material typically is waste, and only a small fraction is an economic commodity—ore. A profitable lead or zinc mine, for instance, might result in 90% or more of the total rock mined being placed in waste storage facilities; a profitable copper mine more than 95%; and a gold or silver underground mine more than 99% after recovering the precious metals. Coal, however, produces much less waste rock from underground mines, especially where the seams contain few waste stringers.

Metals occur in a broader range of host rocks and in more diverse geological settings than does coal. Thus, the geochemistry of metal mines, altogether, is more diverse. Metal deposits differ from coal in other ways. In metal deposits, ore and gangue minerals, host rock lithologies, wall rock alteration minerals, and associated trace metals are more numerous and varied. Metal ores occur in diverse types of igneous, sedimentary, and metamorphic rocks, whereas coal occurs only in sedimentary and metamorphic rocks originally deposited in terrestrial to marine swamp-type settings. Structural and physical characteristics are also more diverse in metal deposits. Whereas coal deposits always occur in stratigraphic layers (or beds), metallic deposits can occur in such layers but also in veins, placers, stockworks, magmatic segregations, disseminations, and a host of other configurations. This greater diversity can require a broader geological background to evaluate mitigation strategies.

Both coal and metallic mineral deposits have commonly associated iron sulfide (pyrite)-bearing units that are disturbed by mining. Pyrite associated with coal is probably related to biological sulfate reduction inherent to the reducing conditions typically encountered in a coal depositional environment. As coal and adjacent layers often contain iron sulfide, so do many metallic mineral deposits and their host rocks. In the eastern U.S. coal beds, for instance, an iron-sulfide-rich unit typically occurs immediately adjacent to the coal seams. In many metallic mineral deposits, a zone (e.g., hydrothermal alteration halo or cupola) occurs adjacent to or closely associated with the deposit, which contains iron sulfides. Such adjacent iron-sulfide-rich layers are difficult or impossible to avoid during mining and can be the major source of MIW. Both types of mining can generate acid and release dissolved metals.

Organic sulfur is believed to be complexed and combined with organic compounds of coal and organically bound within the coal. Typically, this form of sulfur is found in appreciable quantities only in coal beds and in other carbonaceous rocks, not metalliferrous deposits. Generally, the organic sulfur component is not chemically reactive and has little or no direct effect on acid-producing potential.

The mineralogy of metallic deposits is generally more complex than coal deposits, though exceptions do occur. On a ton-for-ton basis, the economically valuable portions of metallic mineral deposits typically comprise only a minor part of a much larger mineralized system wherein many elements and minerals are concentrated. Mines usually are sited only where minerals occur in large enough concentrations and tonnage to be recovered at a profit. Many metallic deposits produce more than one commodity.

Because of the more complex mineralogy, element concentrations, and multiple metals, extractive metallurgy (chemical extraction, hydrometallurgy, beneficiation, smelting, and refining) is much more complex for metals than for coal (Figures 2.1 and 2.2). As a result, the diversity and amounts of recovered minerals and chemical reagents involved in extractive metallurgy

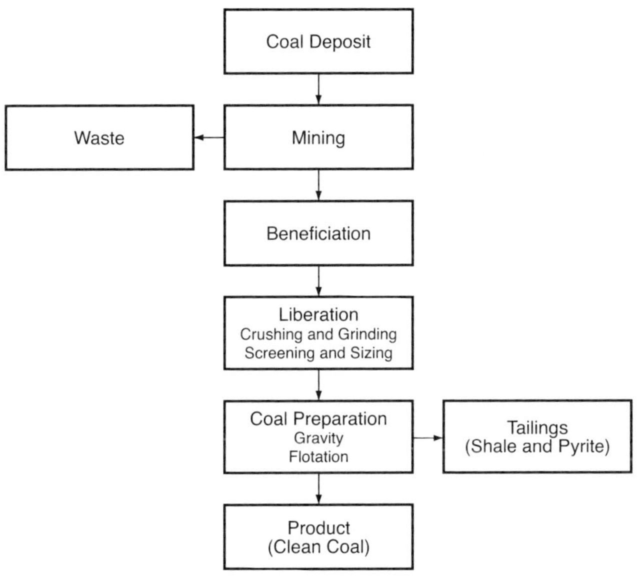

FIGURE 2.1 Major steps involved in coal processing

of metals are far greater than those used in coal extraction, as are the potential environmental impacts from the reagents. Over time, many mines, rock piles, and tailings have been exploited during more than one campaign as miners take advantage of new technological developments in extractive metallurgy and changing societal demands that make lower-grade deposits, waste rock piles, stockpiles, metallurgical wastes, and previously unmined minerals potentially profitable. To produce a clean product for market, coal processing typically requires crushing, grinding, and sizing followed by physical separation of pyrite and shale by gravity or flotation. Polymetallic ores, however, typically require more steps to segregate ore minerals from gangue or from other ore minerals.

FACTORS AFFECTING MIW

As stated by Drever (1988), "The chemistry of most surface waters and shallow groundwaters is the result of interactions between rain or snowmelt and rocks near the earth's surface." Variations in the composition of rain and snowmelt are responsible for some of the differences in near-surface water chemistry. To a significant extent, biological processes and inputs affect water quality. The widely variable compositions in rocks are the principal cause of wide-ranging chemical compositions of surface and near-surface waters. With regard to these water–rock interactions, AD generates perhaps the most profound variations in water chemistry. In addition, AD also promotes rapid water chemistry changes.

Environmental impacts from MIW are mostly controlled by mineral deposit type, mining and metallurgical processing methods, the design of waste storage facilities, and site-specific conditions. Fundamentally, mineral deposits are classified by mineralogy, so the presence of a metal in a certain class of ore deposit and its associated waste rocks will affect the presence of that metal in resultant mine drainage. Copper deposits, for instance, can produce high copper concentrations in the drainage, whereas deposits with low concentrations of copper rarely cause copper to appear in the mine waters. The presence of iron sulfides, which weather to form the soluble

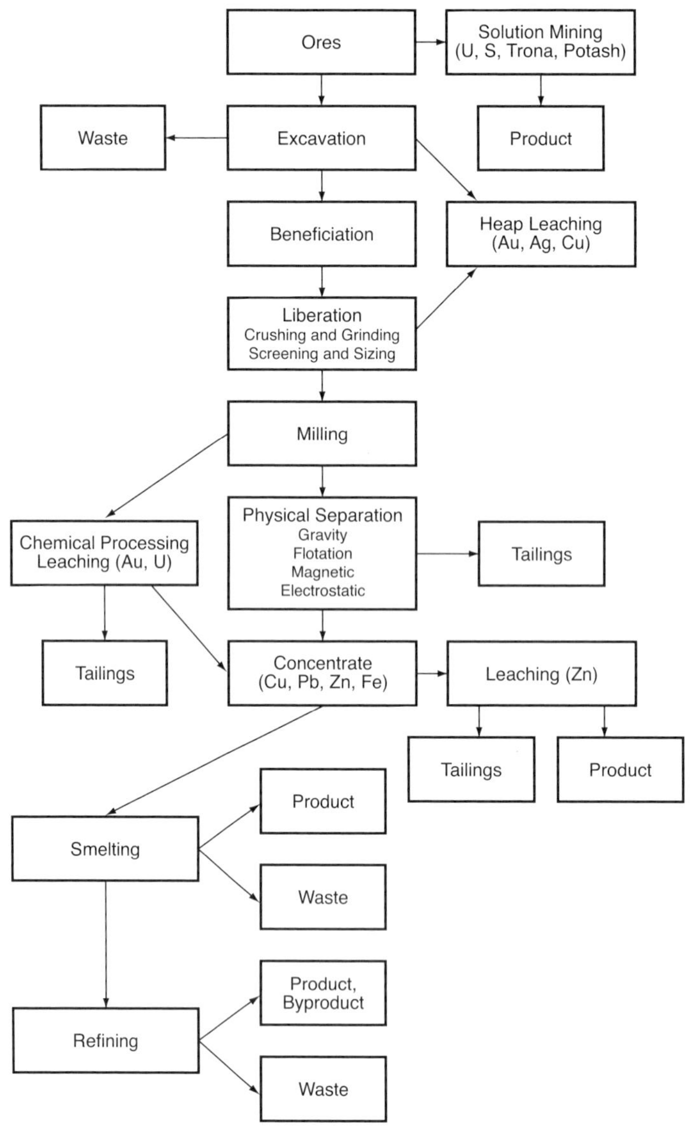

FIGURE 2.2 Major steps involved in extraction metallurgy of metals

iron and sulfuric acid that are characteristic of AD, provides a fundamental control on drainage quality. Once formed, sulfuric acid and ferric iron can react with whatever minerals they contact, thereby releasing metals and other constituents from those minerals in proportion to their rate of dissolution, abundance, accessibility, and reactivity.

Mining and waste storage methods also can affect mine and metallurgical processing drainage quality. Methods that minimize water and rock contact, pyrite oxidation, or biological activity, or that place neutralizing agents (limestone, for instance) in direct contact with iron sulfides all help reduce acid production. On the other hand, methods that ignore these features can promote the formation of sulfuric acid, although deposits containing high amounts of iron sulfide minerals and low amounts of carbonate minerals are generally susceptible to production of AD.

Mining methods (underground, surface, or others, such as in situ or solution mining) are determined by ore-body characteristics and the economics of ore processing. The mining method, in turn, can have different effects on the drainage quality as the method regulates the degree of contact between air, water, and rock. For instance, surface and underground mining locally can expose previously unexposed and unweathered pyritic surfaces through blasting, and that exposure can result in AD. Leaching locally can introduce oxidizing fluids that, to some limited degree, can promote acid generation. Each of these methods affects water quality by changing the degree of contact between potential oxidants and the undisturbed rock. When underground and open pit mines that are below the established water table are flooded with water, acid production ceases due to lack of oxygen.

In summary, the environmental impacts on water and drainage quality are typically related to the specific type of mineral deposit and site-specific conditions. Deposits containing high concentrations of pyrite and other sulfide minerals specifically are susceptible to acid rock drainage (Plumlee 1999). Specific characteristics of the ore body result in specific types of mining (underground, surface, or methods, such as in situ or solution mining) that affect the water and drainage quality in different ways. The complex geochemistry of metallic deposits can result in more diverse ions in solution. The resulting complex water chemistry must be defined prior to remediation.

TYPES OF METALLIC AND INDUSTRIAL MINERALS DEPOSITS

Mineral deposits are classified into types, which are based on features such as host rock composition, mineralogy, and environment of deposition. The deposit types strongly control potential mine drainage quality because they include, in some form or another, the chemical composition of the ore and gangue (Figure 2.3). Although pyrite is a major factor in producing AD and other MIW from metal mines, other sulfide and nonsulfide minerals can play important roles in controlling the generation of acid. Carbonate minerals, particularly calcite and dolomite, can neutralize and buffer pH, and help precipitate metals as hydroxides. Mineralogy is a fundamental component of mineral deposit class.

Numerous classifications have been applied to mineral deposits to aid in exploration and evaluation of mineral resources (Lindgren et al. 1910; Lindgren 1933; Eckstrand 1984; Guilbert and Park 1986; Cox and Singer 1986; Roberts and Sheahan 1988; Sheahan and Cherry 1993), and some of these are useful for predicting drainage quality. Early classifications were based either on the form of the deposit or a combination of form and perceived chemical conditions (Lindgren 1933), while later classifications were based on tectonic settings (Sillitoe 1972, 1981; Guilbert and Park 1986). More modern classifications are based on tectonic setting plus physical and chemical characteristics of the deposits (Cox and Singer 1986; North and McLemore 1986; Roberts and Sheahan 1988; Sheahan and Cherry 1993; McLemore 2001).

In the 1990s, the U.S. Geological Survey incorporated environmental geochemistry into mineral deposit classifications (du Bray 1995; Plumlee 1999). These geo-environmental models, which are defined as a "compilation of geologic, geochemical, geophysical, hydrologic, and engineering information pertaining to environmental behavior of geologically similar mineral deposits prior to mining, and resulting from mining, mineral processing, and smelting" (Plumlee and Nash 1995), are useful for evaluating controls on water chemistry from mine drainage. Plumlee et al. (1994) noted that "a detailed understanding of mineral deposit geology and geochemical processes, which control element dispersion into the environment, is crucial for the effective prediction, mitigation, and remediation of the environmental effects of mineral resource development." In developing geo-environmental models, geologic and geochemical information is

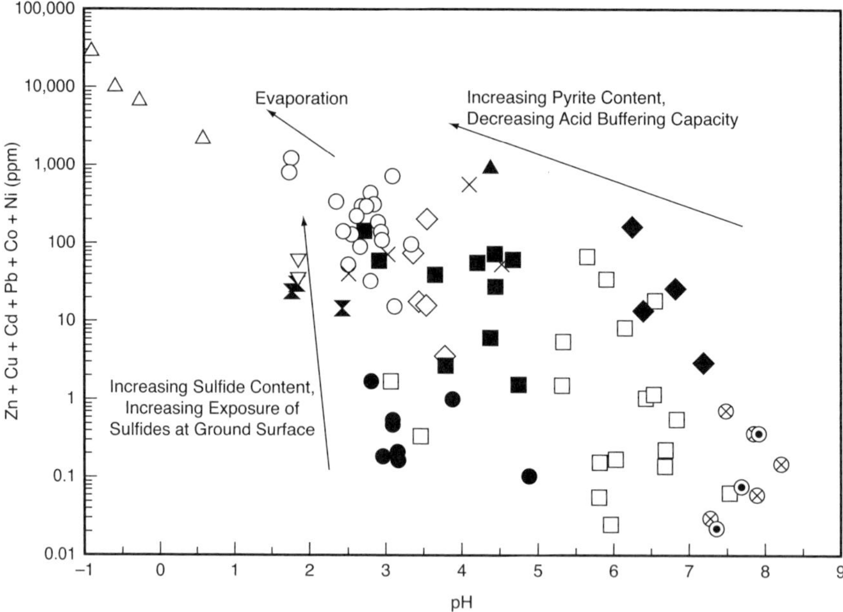

NOTE: Symbols depict waters draining deposits with similar geologic characteristics. All samples were filtered.

Source: Plumlee et al. 1999.

FIGURE 2.3 The Ficklin plot shows pH versus total metal content (the sum of dissolved cadmium, cobalt, copper, nickel, lead, and zinc concentrations) of waters draining from diverse mineral deposits. The purpose of this plot is to illustrate the typical metal content of waters derived from specific types of mineral deposits (Ficklin et al. 1992).

compiled and translated from the language of economic geology and mining engineering to the language of environmental science (Seal et al. 2000; Seal and Foley 2002), and conceptual models have been developed to predict the chemical and physical response to weathering and other environmental processes. Some of the geologic factors that affect drainage quality are summarized in Table 2.1.

Key descriptors in an environmental model are deposit type, related deposit types, deposit size, host rock lithologies, surrounding geologic terrain, wall rock alteration, nature of ore, mining and ore processing methods, deposit trace element geochemistry, primary mineralogy and zonation, secondary mineralogy, soil and sediment signatures, topography, hydrology, drainage signatures, climatic effects, and potential environmental concerns (Plumlee 1999; Plumlee et al. 1999). Determining site characteristics and comparing them with previously determined relationships between similar characteristics and their potential for adverse impacts on water quality can be used initially to assess mineral deposits.

Many types of ore deposits were formed or altered by the convective-driven circulation of water around magmatic intrusions. This process, termed *hydrothermal alteration*, may either add or deplete elemental constituents from the rocks surrounding the intrusive, thereby changing the mineral composition of the rocks contacted by the circulating fluids. Hydrothermal alteration commonly results in the formation of iron sulfides in the rock mass surrounding mineral deposits, although in some deposits the alteration can oxidize the iron sulfides, leaving behind acidic reaction products along with other minerals. Other minerals, including other metal sulfides, clay

TABLE 2.1 Geologic characteristics of mineral deposits that can affect water and drainage quality

Characteristic	Controls	Remarks
Iron sulfide content	Chemical, biological	Oxidation produces acidic waters and also supplies ferric iron, an aggressive oxidant.
Concentration of other sulfides	Chemical	Some generate acid during oxidation, and release of metals during oxidation causes degradation of drainage quality.
Concentration of carbonate, aluminosilicate, and other nonsulfide minerals	Chemical	Many of these minerals can consume acid, and iron and manganese carbonates can produce acidic waters under certain conditions.
Minerals resistant to weathering	Physical, chemical	These control release of metals and drainage quality.
Secondary minerals	Chemical, physical	Soluble secondary minerals can store acid and metals that can be released when the minerals are dissolved.
Premining and pre-erosion weathering and oxidation	Chemical	This reduces the potential for sulfide deposits to generate acid.
Host rock lithology	Chemical, physical	This can consume or generate acid and can increase or decrease resistance to erosion.
Wall rock alteration	Chemical, physical	This can consume or generate acid, can increase or decrease the ability of the rock to transmit groundwater, and can increase or decrease resistance to erosion.
Major and trace elements	Chemical	These can affect metal mobility and drainage quality.
Form of the ore body (vein, disseminated, massive, placer)	Physical	This controls weathering.
Porosity, permeability, and hydraulic conductivity of host rocks	Physical	These control weathering.
Nature and extent of faults, fractures, and joints	Physical	These control weathering.
Deposit grade and size	Physical, chemical	These control magnitude of adverse drainage quality and extent of mining.
Mineral and alteration zoning	Physical, chemical	This can cause variations in environmental signatures, which can affect drainage quality.

Source: Adapted from Plumlee 1999.

minerals, micas, feldspars, epidote, chlorite, and carbonate minerals, can also form in response to hydrothermal alteration. The suite of hydrothermal alteration minerals that form will vary depending on the temperature and pressure of the rocks during the alteration event, although the composition of the mineralizing fluids and mineralogy of the surrounding rock or magmatic intrusive also control the composition of the hydrothermal minerals.

After ore formation, weathering can also affect the mineralogical and chemical composition of the rocks. Weathering is the disintegration of rock by physical, chemical, and/or biological processes that can result in a reduction of grain size, an increase or decrease in cohesion or cementation, and a change in composition of the rock or mine material by solutions and other processes in the rock pile after mining. Ore deposits containing metal sulfides commonly have a weathered portion where mineralized rocks were exposed to groundwater and atmospheric conditions at some point in time since formation.

Hydrothermal alteration is generally more extensive than the economically valuable portions of a mineral deposit. Mining the salable commodities of mineral deposits commonly requires excavation of waste rock adjacent to the valuable minerals, and such excavations typically take place in hydrothermal alteration zones. Commonly, those waste rocks can contain iron sulfides, especially pyrite.

The altered country rocks vary in extent, from a few millimeters to many kilometers from the vein or mineral deposit, and they form as alteration zones, halos, envelopes, and selvages on veins (Guilbert and Park 1986). Alteration zones or halos are typically more widespread than the actual mineral deposits. Pyrite and marcasite are common alteration minerals and are responsible for naturally acidic water in some mining areas (Plumlee 1999; Plumlee et al. 1999). Clay alteration, which is common in many mineralized areas, can actually consume acidity under some circumstances, although the rate of consumption in natural systems is perhaps not as great as laboratory tests indicate.

CHAPTER 3

Types of Metal Mining and Extractive Metallurgical Processes

METAL MINING

Because ore bodies come in many sizes and geometric shapes, mining methods must conform to each specific ore body and metal recovery economics. Mining can be classified into four basic methods:

1. Surface,
2. Underground,
3. Placer, and
4. Solution.

Summaries of the types of mining can be found in works by Stout (1967, 1980), Thomas (1973), Hartman (1992), EPA (1995 1997), Gertsch and Bullock (1998), and Whyte and Danielson (1998). A brief description of each method follows.

Surface Mining

Minerals located close to or at the surface are extracted by removing soil and waste rock overlying, adjacent to, or intermixed with minerals of the deposit. Typically, surface mining involves overburden removal, blasting, mucking (picking up), loading and hauling, and dumping. Typically, backfilling is too expensive, but where mine configurations, environmental benefit, and cost allow, some or the entire mine can be backfilled. There are three types of surface mining: strip, open pit, and quarry.

Strip mining. An overburden layer is removed or stripped to expose an underlying layer of ore or industrial mineral. In a typical strip mining operation, a single strip of overburden will be removed and placed aside, exposing the underlying ore layer for mining. Once the exposed strip of ore layer is removed, a second strip of overburden will be removed and placed in the channel left by the first strip. This process will continue until the entire layer is mined or until overburden thickness becomes too great to allow profitable mining. Strip mining is employed where the overburden is too thin to form a sturdy mine roof or where stripping is cheaper than underground mining.

Open pit mining. Where the shape of the ore body will not accommodate strip mining and the ore is close enough to the surface to be mined at a profit from an open pit, successive layers of rock (benches) from the surface to depth are systematically removed, thus forming an open bowl or pit. The layers can contain either waste rock or ore but usually contain both. The shape of the ore body and the grade of ore in each successively deeper layer and the mechanical characteristics of the rocks in place determine the shape of the pit. The depth of the pit is determined by both the depth of the ore and the relative cost of mining versus the value of recovering additional ore at depth and also depends on environmental cost factors, such as engineered containment of

FIGURE 3.1 Aggregate quarry near Deming, New Mexico

potentially acid-generating ores and waste rock. Typically, some unmined portion of the ore body is left in the pit floor, because the relative cost of removing waste rock exceeds the value of recovering additional ore at depth.

Surface mining typically alters the flow of surface and groundwater by either restricting existing flow paths or introducing new ones. Such mining can change the composition of waters by inducing interactions between water and minerals in the ore body or associated alteration minerals or waste rock that is excavated and hauled to waste storage facilities during mining. If surface mines are dewatered to accommodate mining, groundwater composition can be altered because of exposure of sulfidic minerals that, prior to mining, lay below the water table and generally below the zone of oxidation.

Quarry mining. Though somewhat nonspecific, the term *quarry* is used typically to differentiate an open pit metal mine from an open pit mine that recovers either aggregate or dimension stone (Figure 3.1). The term also is applied (incorrectly) to a few aggregate or dimension stone operations, which, though located partially or wholly underground, are accessed through horizontal rooms. Aggregate and dimension stone mining generally will not be discussed in this handbook.

Underground Mining

Such mines are either too far below the surface to be accessed by either an open pit or strip mine or are more profitably mined by underground methods, which leave a roof (or back) or rock above the mining levels; they may be an extension of an open pit mine. Underground mining (USBM 1994), can be characterized as the following:

- Methods producing natural or minimal support (room-and-pillar),
- Caving methods where failure of the back (roof or overburden) is required (sublevel or block caving), and
- Methods that require substantial artificial support (cut-and-fill or stope backfill).

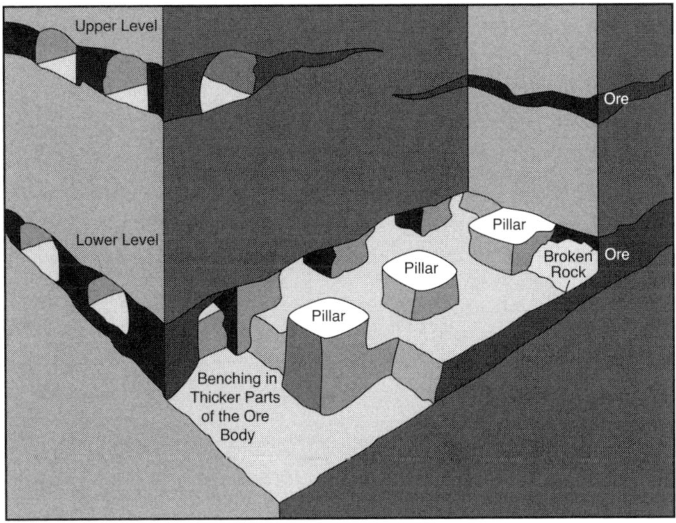

FIGURE 3.2 Room-and-pillar mining

Room-and-pillar mining . A series of rooms are excavated into the deposit, leaving pillars or columns of material (Figure 3.2), in many cases ore, to help support the mine roof (USBM 1994; Farmer 1999). Some coal deposits are mined by conventional room-and-pillar mining, but underground room-and-pillar methods for many metal deposits differ from those of coal primarily by the equipment used, ore-breaking methods, and ore removal from the mine. Whereas coal can be cut continuously with either a shearer or continuous mining machine, most metal mines use pneumatic, jackleg, or hydraulic percussion drills to drill small-diameter holes, which are loaded with blasting media to break the rock.

Caving mining. This method requires that the ground be weak and permanently available for closure and isolation so that it will break into a fractured mass, leaving few options for land use after mine closure. *Sublevel caving* is used in wide, steeply dipping, vein-type or tabular ore bodies. Commonly, ramps and large boreholes are used to access sublevels, which are spaced 6.1 to 12.2 m (20 to 40 ft) apart vertically and extend through the ore body. Mining occurs sequentially in block segments 61 to 152 m (200 to 500 ft) on a side in all three directions. Haulage drifts (tunnels with one entrance) are dug under the ore body. At approximately 30-m (100-ft) intervals along each drift, generally in a checkerboard pattern, 3 to 6 m (10 to 20 ft) in diameter, raises connect the drifts with another series of cross drifts. The section of ore to be caved is drilled in a fan pattern from the bottom, and blasting is used to break up the ore for removal at the bottom of the block or slice through draw points. The ore continues to fall from the bottom of the blasted block as it is pulled from the draw points, for it now has no foundation (Figure 3.3). Entry is precluded once a block begins to cave in.

Block caving (Figure 3.4) is similar to sublevel caving in that it utilizes gravitational forces to move the ore. This method is used to mine large ore bodies that have relatively consistent grade throughout and consist of rock that will break into manageable sized fragments (Julin 1992). Because of its weight, the ore is allowed to collapse in a controlled fashion into chutes. After a horizontal slice of ore is removed from the base of a rectangular section of the ore body, the unsupported block of ore breaks up and caves in under its own weight. The broken ore is drawn off from below through draw points as the caved ore falls because of gravity.

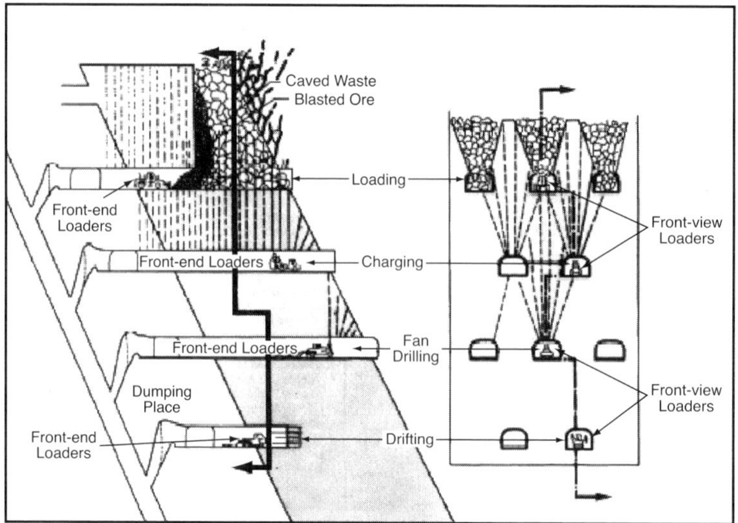

Source: Kvapil 1992.

FIGURE 3.3 Sublevel caving

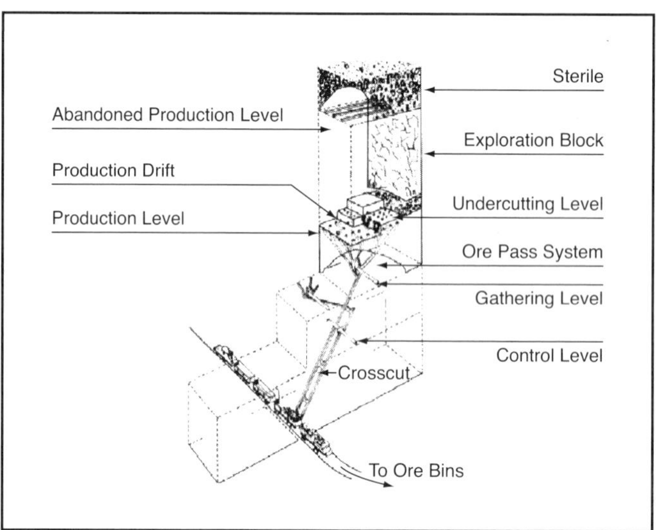

Source: Julin 1992.

FIGURE 3.4 Block caving at El Teniente, Chile

Cut-and-fill mining. When ore deposits are irregular and tabular, with varying thickness, inclination, and strength, the cut-and-fill mining method (Figure 3.5) can be employed to accommodate these variables. After the ore is removed from a small segment of the ore body (20 to 4,000 ft^3), waste rock is typically mixed with cement and placed back in the excavation for support. This process is repeated, advancing along a horizontal level through the ore body. After

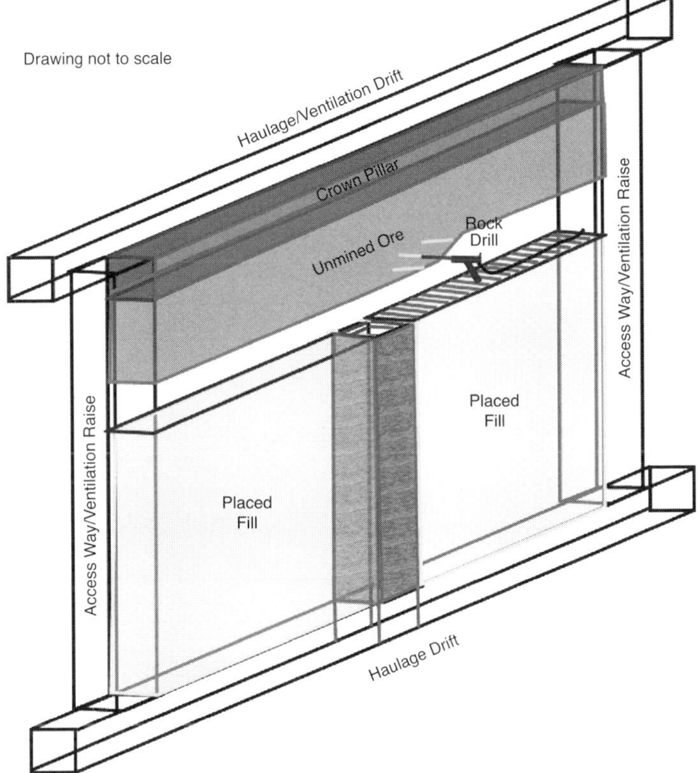

NOTES: Fill access ways, ventilation fans, ladders, equipment, and miners omitted for clarity. Distance between access ways/ventilation raises varies from 200 m to 300 m; haulage/ventilation drifts about 122 m apart vertically.

FIGURE 3.5 Cut-and-fill method

a slice of ore has been removed and backfilled, mining continues upward (in overhand mining) or downward (in underhand mining) and another slice is started.

In all its forms, underground mining usually involves blasting, stoping, mucking, hauling, and skipping (vertical haulage) of ore and waste to the surface. Backfilling of waste is practiced at some underground mines, which can result in waste rock being consumed underground. Surface waste rock or tailings can also be transported for use in the mine. Unless waste rock is disposed over significant-sized areas, underground mining typically does not create nearly as much surface disturbance as surface mining because a relatively smaller amount of waste material is brought to the surface. However, surface subsidence can develop where subsurface mining is shallow or where mining occurs beneath weak rocks. Surface facilities can be extensive for both types of mining, depending on the type of mineral processing and whether underground milling or extraction is employed.

Like surface mining, underground mining can alter the groundwater flow. In addition, the exposure of sulfidic rock to oxygen can change the composition of the mine water, through disaggregation of the rock in place by blasting or tension fracturing that follows underground excavation or through mine dewatering. At many older underground mines located in steep terrains, such as in Colorado, underground mining was locally made possible through construction of a local adit (horizontal shaft) that drained groundwater to a nearby stream. Such draining adits, which were constructed miles underground, could dewater many square miles of mine workings and potentially provide water for other uses. However, if left untreated, these draining adits pose additional environmental problems by allowing MIW to dominate local streams and rivers.

Placer Mining

This variation of surface mining involves the removal of natural concentrations of heavy minerals, such as gold, tungsten, tin, zircon, or apatite, from unconsolidated sediment or soil by gravity processing. Drilling and blasting are generally not required for most placer mining. Where blasting is required, it is usually minimal and is intended to either access valuable minerals that are locked in locally cemented layers or remove rock impediments around which placer sediments have collected.

Placers occur in alluvial fans, colluvial sediment, bench or terrace gravel, point bars, or meander bars in streams. Residual placers, also called lag deposits, are formed directly on top of veins or other mineral deposits. Commonly, lag deposits drift downhill, incorporating soil and rock debris. Placer deposits are composed of high-density minerals, such as gold, that are more resistant to dissolution by precipitation and shallow groundwater than their surrounding host minerals.

During fluvial (river flow) events, sediment that contains heavy metals is transported and deposited in alluvial and stream deposits. Heavy minerals concentrate by gravity in stream channels, in valleys, and in alluvial fans. If the heavy mineral deposits are concentrated and large enough, they can present mining targets.

Traditionally, placers were mined by gravity-separation techniques such as panning, sluicing, and dredging. Miners would divert streams, sending smaller streams aside and leaving stream beds exposed. However, some placers in California, Colorado, Nevada, Alaska, Saudi Arabia, and elsewhere are mined by underground methods.

Hydraulic mining, a variant of placer mining, was preferred for some placer deposits, especially gold. In hydraulic mining, water from a high-pressure hose is used to mine unconsolidated sediments and weathered bedrock. The high pressure is often created by connecting a water source at a high elevation to a large-diameter loading end of a hose leading to a smaller-diameter exit end of the hose. The weight of the column of elevated water in the hose produces a high-pressure water jet that is used to cut into a hillside or alluvial deposit to loosen ore-bearing gravels, soil, and weathered bedrock and wash them into a sluice box or other similar type of gravity-separation device to capture high-density metals, such as gold (Figure 3.6). This method of mining, due to the scale of production and speed of extraction, can have vast environmental impacts, including sedimentation and erosion (Pentz and Kostaschuk 1999). In Alaska and other cold areas, the local gravel was thawed by steam or wood burning. Although acid drainage is rarely a concern in placer mining, other drainage quality concerns can develop, such as sedimentation and mercury amalgamation.

Solution Mining

This form of mining involves the circulation of leaching solutions directly into ore zones and recovery of the leachates for processing through a series of injection and recovery wells. Thus, the ore is processed in its original geologic location, negating the need for excavation and transport to a processing facility, as is typical of conventional mining. Solution mining, which includes in situ leaching (Schlitt 1999), can be applied to extract solids that are easily leached, dissolved, or melted. It is distinct from mining of solutions containing recoverable concentrations of metals or

FIGURE 3.6 Hydraulic mining at Elizabethtown-Baldy District, New Mexico, in the late 1880s

other products and also different than heap, vat, and stirred tank leaching, which are metallurgical processes carried out in processing facilities. Although in situ solution mining may involve the installation of injection and recovery wells similar to those used in oil and gas production, it is a distinct process. Several commodities, including uranium, copper, manganese, salt, potash, nahcolite, and sulfur, have been mined economically using solution mining techniques.

In practice, leaching solutions may be acidic, neutral, or alkaline water, either heated or unheated. In deposits where ore-bearing minerals are sparingly soluble, oxidants such as dissolved oxygen or hydrogen peroxide and complexing agents such as chlorine dioxide gas, or $CO_2(g)$, are often used to accelerate leaching rates. In deposits where soluble mineral forms are solution mined, such as soda-bearing minerals like nahcolite, heated water is often sufficient for extraction. In either type of deposit, the host rock must be permeable enough to allow injected fluids to contact the ore-bearing minerals and also to be recovered again through boreholes. If the host rocks are not sufficiently permeable, they can be fractured by blasting from boreholes. Dissolution of an entire mineral-bearing zone or of all the ore minerals within a mineralized zone is not always accomplished. In some cases, total dissolution of the mineral zone is deliberately avoided to prevent collapse or surface subsidence.

In situ operations, especially those that mine uranium or copper, typically consist of a pattern of injection and recovery wells. Leach solutions (lixiviants) for dissolving uranium are usually alkaline with high concentrations of carbonate and oxidants of either dissolved oxygen or hydrogen peroxide to promote leaching of uranium oxide minerals. Leach solutions for copper are often acidic with added chemical reagents. The lixiviant is injected into the ore zone, and, as it migrates through the permeable host rock, contacts the ore minerals, oxidizes them, and leaches the metallic minerals from the host rocks. The metal-rich leachate is then pumped to the surface where metals are recovered by solvent extraction, ion exchange, cementation, or other processes.

Another solution mining technique, the Frasch process, was used to extract sulfur, especially from salt domes in Texas, Louisiana, and the Gulf of Mexico and from bedded sulfur deposits in west Texas. In the Frasch process, wells were drilled into native sulfur-bearing formations. The well pipe was lined with two additional pipes to create three concentric pipes, and superheated water was pumped into the outermost pipe. Air was compressed into the center pipe to force a mixture of elemental sulfur, hot water, and air up through the middle pipe. In other areas, a field of hot water injection wells was constructed adjacent to separate solute withdrawal wells. Superheated water, above the temperature at which sulfur melts (approximately 116°C), was pumped under pressure into the wells. As the sulfur melted, a mixture of liquid sulfur and water was pumped to the surface by applying air pressure through the central pipe. After recovery at the surface, the solution was discharged into tanks, and the sulfur cooled into a solid mass and stockpiled until shipped.

Some halite (salt, NaCl) deposits are mined using solution mining techniques, either to recover halite for sale or to create underground cavities to store material, such as crude oil or natural gas. In these cases, locally derived fresh or salt-undersaturated water is injected into the salt body through a central pipe. The solution is then withdrawn from the subsurface and is either evaporated on the surface to recover the solid salt or reinjected into deep saline aquifers for permanent disposal. Whereas Frasch sulfur solution mining removed only sulfur from the native ground, halite solution mining removes virtually all of the host rock, as the host rock is the salt formation itself. Halite solution mining differs from Frasch mining in that halite does not require heated water for dissolution.

In northwestern Colorado, nahcolite (sodium bicarbonate, $NaHCO_3$) is mined by solution mining techniques from shale formations in the Piceance Basin. Hot water is injected into nahcolite-bearing formations to dissolve nahcolite from the host rocks, and the resulting $NaHCO_3$-enriched solution is pumped to the surface for cooling and recovery of the solid sodium bicarbonate. Voids created by the nahcolite dissolution can cause strata above the solution-mined formation to collapse, so cavity diameter and height are controlled to prevent contamination of freshwater aquifers above the mined zones and to control surface subsidence.

EXTRACTIVE METALLURGY

Mechanical and chemical processes are used to extract metals from ores. Though some industrial minerals contain a target mineral (e.g., halite), most metallic ore bodies comprise a suite of useful or valuable ore minerals that are tightly mixed with noneconomic gangue minerals. The amount of material handling required and the chemical and physical complexity of metallurgical processing depend on the specific physical and chemical complexities of each ore and gangue assemblage. There are basically six steps in extractive metallurgy:

1. Comminution (size reduction),
2. Liberation,
3. Leaching,
4. Concentration,
5. Smelting, and
6. Refining.

The type of metallurgical processing required depends on the mineralogy of the ore and the metals to be recovered and sold. Extractive metallurgy techniques are summarized by Biswas and Davenport (1992), Gilchrist (1989), Lewis (1983), Mackey and Prengaman (1990), Gill (1991), Leonard and Hardinge (1991), Marsden and House (2006), and Wills (1992).

Following excavation from the host rock, ores are sometimes crushed and ground to sand-sized particles (more or less) to help segregate ore minerals from gangue minerals for further purification. One or more concentrates of ore minerals can be separated and leached or physically recovered before being produced and shipped to be smelted and refined. Alternatively, ores can be crushed to roughly cobble size and stacked on a lined pad for chemical reagent heap leaching or placed on a lined pad for run-of-mine chemical reagent dump leaching. Because processing solutions and mineral wastes from metallurgical processing can affect drainage quality, they must be properly managed.

Ore comminution is the process "in which the particle size of the ore is progressively reduced until the clean particles of mineral can be separated by such methods as are available" (Wills 1992). Particle size reduction is typically followed by *sizing* and *classification* to segregate liberated particles into size classes for the next stage of processing (liberation), whereas coarser particles may be treated to additional crushing or grinding. Mineral segregation techniques, including leaching, take advantage of sometimes subtle physical and chemical differences between mineral species, including mineral size; mineral shape; density (specific gravity); wetting characteristics (hydrophobic vs. hydrophilic mineral surfaces); magnetic susceptibility; induced magnetism; electrical conductivity; attraction to or repulsion from surfactants, coatings, and surface treatment chemicals; and melting temperature. Mineral *separation* techniques, such as gravity and flotation, usually follow to produce concentrates of certain minerals and native metals. Once the ore has been concentrated, the concentrates are sent to a smelter for further purification. Refining is required after smelting of some concentrates in order to produce a near-pure product.

Mill-processing facilities range from those housing relatively simple gravity-separation equipment to those with roasting, autoclave, flotation, leach processing, or other processing equipment. Knowledge of milling and metallurgical processing technologies, particularly chemical reagent treatments, is useful for evaluating environmental mitigation that can be necessary at active, inactive, and abandoned metallurgical processing sites.

Significant volumes of crushed and ground material, stockpiles, tailings, and so forth can remain after milling and metallurgical processing. The volume of tailings to be stored varies with the grade of the ore and is generally a significant percentage of the total rock mined. After metallurgical processing, sulfidic ore deposits can produce tailings that are potentially acid generating.

At each step in the process from the mine to the mill, concentrator, smelter, and refinery, a waste material is generally produced. Tailings, smelter slag, and processing solutions can be sources of AD or other drainage quality impacts. Disposal facilities design at each step should consider potential environmental effects on MIW, from the weathering of the ore, concentrates, smelter slag, tailings, and so forth.

Mining for heap and dump leach processing involves either conventional open pit or underground mining, although the relatively high cost of underground mining usually justifies the milling of the ore to achieve higher or quicker metal recoveries. After the ore is mined, broken up, stacked on a lined heap or dump pad, a leaching solution is applied to the heap or dump, which dissolves the metals. The metal-bearing solution is recovered by gravity, contained by the liner, and processed to recover the metals of interest. Heap and dump leaching are types of hydrometallurgical processing. Process waters must be managed to prevent adverse effects on drainage quality.

CHAPTER 4

Phases of Mining

One of the principles of sustainable development in mining is that every mining or processing operation has a beginning and an end (Figure 4.1; MMSD 2002). The mine life cycle can be divided into four stages:

1. Exploration,
2. Development,
3. Operations, and
4. Closure and postclosure.

Figure 4.1 briefly describes the mine life cycle.

EXPLORATION (PREMINING OR UNDISTURBED)

The first step of the mine life cycle is prospecting for mineral deposits (Hartman 1992). Once a mineralized area is located, the project phase follows, which determines whether the mineralized area contains enough potential ore to be mined economically. If a potential mineral deposit is located, then operators typically conduct feasibility studies to determine whether the deposit can be mined and processed economically. Claim staking, permitting, road and trail building, and sometimes helicopter transport occur during the exploration and project phase, and can continue into the advanced project and feasibility phases.

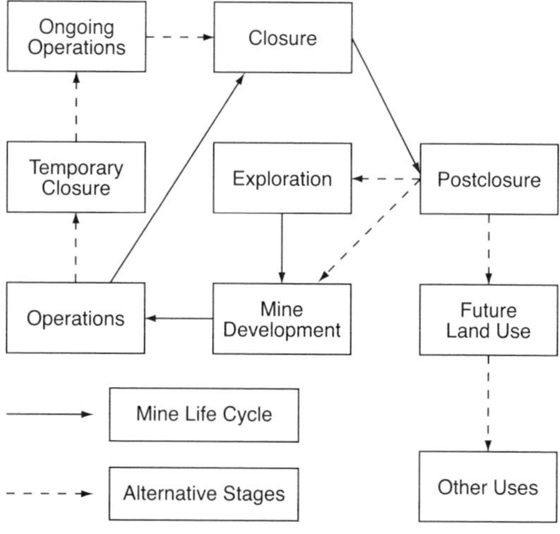

Courtesy of Dirk J.A. Van Zyl.

FIGURE 4.1 Mine life cycle stages

A variety of exploration techniques can be employed, including literature searches, geological mapping, geochemical exploration (rock and soil sampling and analysis), geobotanical surveys, geophysical and remote sensing surveys (surface, subsurface, airborne, and satellite), aerial photography, and drilling. Although most exploration programs never result in operational mines, these surveys can be useful in identifying, predicting, and mitigating AD and other MIW and defining natural levels of background concentrations of metals and solutes in soils, rocks, groundwater, and surface water.

Mineral exploration is typically expensive and can require 10 years or more of study before a decision is made to develop, sell, or reclaim the deposit. Today any areas disturbed during exploration must be reclaimed. Impacts can include habitat disruption or destruction, road construction, exposure of acid-generating rock, cross-aquifer mingling, pollution from fuel caches, and more. Modern exploration and development of potential ore deposits require reclamation of trench and drill sites and roads after subsurface rocks are sampled, if no further development is planned. Erosion from trail and road building and trenching locally can impact water and drainage quality. Acid drainage in some areas can result from exposed outcrops of sulfide-bearing rock.

Historically, prospectors sampled at the surface and panned in streams to identify a potential area for continued exploration. Once a prospective area was identified, the prospector would dig pits, shafts, or adits to test the mineral occurrence and determine the grade and extent of the deposit. Many areas today contain numerous prospect or exploration pits, shafts, or adits that never yielded any mineral production and were not reclaimed. These sites can adversely affect the drainage quality, which can include AD.

Modern explorationists, typically geologists, have more tools for locating ore deposits than their predecessors. Satellite images and other remote-sensing imagery, geologic maps, geophysical studies, and geologic reports can be examined to identify favorable areas for exploration. Often, some of these same techniques can be used in environmental studies. Once mineralized areas are identified, the explorationist can conduct field examinations of the area. Surface sampling, test pits, drilling and trenching, bulk sampling, and even panning can be used in the beginning phases of exploration. Outcrops are examined for favorable lithology and mineralogy, and samples are submitted to laboratories for chemical analyses. Most samples are solid (i.e., rocks, soils, stream sediments); other sample media can be collected and chemically analyzed (i.e., biological, water, air). The explorationist can also use geophysical and geochemical techniques. Detailed geologic maps are prepared at many mines to help characterize the rocks at the surface and to predict the rocks in the subsurface. Despite the numerous exploration techniques available today, ultimately the deposit typically must be drilled. Several drilling techniques are available. In some cases, test (or exploration) pits, adits, and drifts are excavated to provide material for metallurgical testing and other purposes.

Environmental evaluations normally are used in site characterization and feasibility studies to determine whether the deposit is economic when the costs of waste containment and closure are included. Land use restrictions, such as proximity to wilderness areas, national and state parks, and other areas can be taken into account in the feasibility of developing an ore deposit. Regulations governing exploration programs vary from state to state and country to country. Some exploration projects are so extensive that detailed operational and reclamation plans must be developed. Permitting is required for exploration in most areas in the United States and elsewhere.

DEVELOPMENT

If an economic ore deposit is found, the mine life cycle can proceed to the second stage: the development of the mine. One of the most important parameters that must be determined during the development stage is characterization and definition of the site-specific baseline and background conditions. Exploration can continue throughout the mine life cycle to expand reserves and mine life. Plans for permitting, operations, waste containment, and mine closure are generally developed at this stage, with requirements varying from site to site.

Site development typically includes reserve drilling and the construction of mine and haul roads, rail lines, terminals, and utilities; mine, metallurgical processing and waste containment facilities; and support and housing facilities. Metallurgical processing development normally includes metallurgical laboratory and pilot-plant operations, as well as construction of metallurgical processing facilities (mills, pipelines, conveyance facilities, leach pads, processing ponds, water treatment facilities, tailings storage facilities, and so forth). Where discharge of MIW will need to be managed during active operations, pilot testing of water treatment processes at the site may be required.

OPERATIONS

Mine production includes extraction of ore and waste rock by blasting, mucking (removing the broken material from the mine), haulage, ore stockpiling, and storage or reuse of mine wastes, followed by metallurgical processing of the ores. Dewatering of the deposit can occur during development, which continues as the ore is mined, and may need to be maintained throughout the life of the mine if inflows continue. If there is excess water from the operation that needs to be discharged, mine dewatering effluent can require water treatment. The operational units of mining can include the mine, overburden, waste rock, and stockpiles, whereas metallurgical processing can include crushing, milling, leaching, and tailings storage. Finely ground waste from metal extraction, called tailings, are managed in engineered facilities.

Environmental effects of mineral production can include impacts to wildlife and fisheries, including habitat loss; changes in local water balance; sedimentation; release of chemical reagents and MIW into tailings ponds or leaching solutions; formation of unstable dams and highwalls; impacts related to construction of tailings ponds or leaching pads; potential acid generation from ore, waste rock, pit walls, and tailings; metal leaching by acid and alkaline drainage; and wind-borne dust. Most of these impacts are routinely resolved as part of the development and production stages.

Smelting and refining normally produce slag and dust wastes. Sulfur dioxide and smelter dust can be produced and must be managed to prevent impacts on air, industrial hygiene, and drainage quality. Sulfuric acid is sometimes recovered and sold as a by-product of sulfide ore and concentrate smelting or roaster operations.

Milling, smelting, or preliminary refining can produce a concentrate or alloy, which must be further refined to produce purified metals. Although it was not always the case, metal smelters and refineries today have become specialized facilities that may not be profitable to operate at all mining sites and are sometimes located in populated areas away from mining districts so that they can receive and process concentrates from a number of metallurgical operations. Refining of the recovered metal can involve smelting, chemical, and electrochemical processing, which must be properly managed to prevent drainage or air quality impacts.

Most mines that temporarily cease production, because of labor problems, economics, natural disasters, or other reasons, still require ongoing monitoring and drainage management during the standby phase.

CLOSURE AND POSTCLOSURE

Mines and metallurgical facilities that have plans based on life-cycle considerations will develop a mine closure plan, prior to operation, that accounts for closure costs. Closure plans must be revised during production to accommodate changes in mining and processing, but generally these should be slight if proper upfront planning was established and the expansion during operation is not significant. If mining, metallurgical processing, or water quality discharge regulations change during mining, closure plans will likely need to be revised. Depending on local regulations and landowner requirements, closure can involve recontouring pit walls and mine waste dumps, covering reactive tailings and waste dumps, decommissioning roads, dismantling buildings, reseeding and planting disturbed areas, ongoing monitoring, and possibly water treatment for an extended period after closure of other production facilities.

Problems that can appear after mining and metallurgical processing include erosion, pit wall and other highwall instability, surface subsidence, storm water management, and seepage of MIW into ground and surface waters. When pit dewatering operations cease, pit lakes can form as the level of the groundwater rebounds (Figures 4.2 and 4.3). These types of problems can locally affect wildlife, fisheries, habitat, vegetation, air quality, and agricultural or drinking water supplies, and can generate physical hazards to humans and wildlife if not properly managed. Plugging of shafts and adits needs to be carefully evaluated, because plugging can cause changes in the groundwater flow and impact the groundwater in some cases.

Closure plans typically establish a postmining use of the site commensurate with the landowner's desire and adjacent local land uses. Some sites will require long-term monitoring. In concert with the plans for final closure, mines and metallurgical processes can operate from exploration through all phases of production.

HISTORIC AND ABANDONED MINES

Historic and abandoned mines can offer challenges to mining, environmental, and community interests. These mines, some of which are quite old, can have owners with insufficient financial resources to reclaim the properties or who can be bankrupt. In some cases, the land reverted to government agencies as a result of inactivity on the property or nonpayment of property taxes. These mines are rarely profitable to operate because of economic conditions, including metals prices; flooding of works; obsolete technology; and modern environmental, health, and safety standards. However, most historic mine sites in the western United States pose no safety or environmental risk.

Before attempting reclamation of historic mines, it is generally useful to know the history of exploration, development, and production. Mines that operated prior to the enactment of modern environmental laws, which were initiated around the early 1970s, had different operating philosophies and knowledge than mines of today. Many of these mines were under financial constraints or wanted to make a profit or produce critical metals quickly and had limited access to efficient extraction technologies. Many operations mined and processed only high-grade ores and left the lower-grade material.

Exploration facilities, which commonly involved construction of pits, shafts, and adits that were not necessarily utilized in ore production, were often left abandoned and unreclaimed. Processing facilities (mills, tailings, stockpiles, etc.) generally were built near the production or haul shaft or portals to minimize haulage without consideration or knowledge of eventual erosion, water quality problems, and ecological restoration. Processing wastes were often dumped near the processing unit, in many places into local drainages, without reusing the processing water

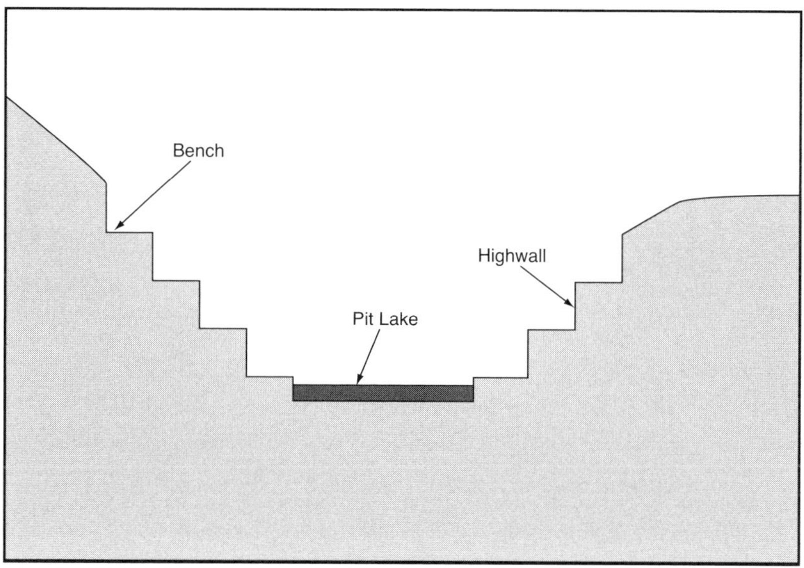

FIGURE 4.2 Sketch of open pit mining approaching closure. A pit lake will develop if mining proceeds below the groundwater table and otherwise remains undrained.

FIGURE 4.3 Chino porphyry copper mine, Santa Rita District, Grant County, New Mexico

FIGURE 4.4 Alamitos tailings pond remains from processing copper, lead, zinc, gold, and silver from the Pecos massive sulfide deposit in the 1930s, Santa Fe County, New Mexico. This tailings pile has since been covered and seeded.

(Figure 4.4). As a result of these practices, instability and erosion are common at historic and abandoned mines.

When society was motivated by Industrial Age or Manifest Destiny attitudes, historic metal mines were developed by underground and surface mining methods that were as inexpensive as possible and were operated without environmental consideration of more modern times. In historical times (pre-1980s), worker safety, postmining safety, and mine stability were not priorities. Historic mines typically have few ventilation shafts and can have poor or nil air circulation. At many mines, mine surveying and mapping typically were not carried out or were minimal, and documentation has been lost. Upon depletion of the ore, older mines were abandoned, and mine waste and tailings were left as they were created—without caps or vegetative cover.

In many mining districts, drainage quality was adversely affected, but because typically no baseline data exist on conditions prior to mining, the cause and extent of the drainage quality impacts must be inferred using geological, geomorphological, and hydrological principles and observations. Except for the presence of non-native reagents, it is generally impossible to differentiate natural drainage conditions from those caused by mining and metallurgical processing. Information to help establish a baseline can come from historical data; published geology of unmined and mineralized areas; water analyses from monitoring wells; leaching studies; statistical analyses; isotopic studies; mapping and sampling of similar geologic, biologic, and hydrologic surrounding areas; identification of background by subtracting mining influences; and computer modeling. Although baseline conditions can be used to establish remediation goals, studies to establish historic conditions nearly always have uncertainty.

CHAPTER 5

Factors that Control Environmental Impacts from Metal Mining and Metallurgical Materials

Although many things can affect environmental impacts from metal mining and metallurgical materials, water is predominant. The amount of water is controlled by the climate and the hydrologic cycle. Before any sampling, modeling, remediation, and monitoring occurs, the chemistry of water before, during, and after formation of acidic fluids must be understood. MIW is controlled by the forms of sulfur in the rock, and the sources of the acid drainage and the transport of the constituents in the water can dramatically affect the environment. Finally, there are specific practices in metal mining and processing, such as cyanide, that affect MIW.

IMPORTANCE OF CLIMATE

Climate controls the amount of water available to produce acid drainage and MIW. Although acid drainage can easily form in either wet or arid climates in mineralized areas containing pyrite (or other acid-generating sulfide minerals), any effect of AD on drainage quality can be critical in arid climates or where dilution water from rainfall is limited. In areas with strong seasonal climatic patterns, such as monsoons, AD can show a seasonal pattern, whereas areas with steady year-round rainfall do not always show any specific pattern with time. Overall, climate plays a role in all aspects of remediation, sampling and monitoring, water quality prediction, MIW prevention, environmental impact mitigation, geochemical and hydrological modeling, and formation of pit lakes (all of which are addressed in separate handbooks in the series).

A proper remediation plan accounts for the possibility of flooding and the potential effects. In some areas, the effects of extreme temperature and freeze–thaw effects are important (i.e., high mountain terrains, Arctic tundra areas). It also is important to remember that at some time in the future some areas can become wetlands for a period of time. Some mineral occurrences are in areas so topographically diverse that local climatic conditions can be extreme. For instance, snow can occur in higher elevations, while rain appears in the intermediate elevations, and nil precipitation falls in the lowest elevations. Because of extreme winter conditions, some areas are accessible for only a few months of the year, and the lack of accessibility impedes remediation. Growing seasons vary according to elevations, which affect the ability to reclaim a site. In summary, the specific climatic conditions for each area must be determined prior to any reclamation for that plan to be successful.

HYDROLOGIC CYCLE

The hydrologic cycle, which describes the regional movement of water, is an important model that enables us to understand how climate and weather relate and to explain local weather and water availability phenomena (Figure 5.1). Simply stated, water from the ocean and land evaporates to form clouds, precipitates over land or oceans, and eventually flows back to the ocean as an endless cycle. Once water reaches land, it follows several pathways to the sea. Several major inflow and outflow components to the hydrologic cycle are shown in Table 5.1.

Solar radiation controls part of the hydrologic cycle. As the sun heats the earth's surface (both land and ocean), water evaporates (changes from liquid to gas), forming clouds. Cooling in the upper atmosphere causes condensation (changes from gas to liquid) and precipitation (rain, snow, sleet, or hail). Water in the soil can be evaporated into the atmosphere or absorbed by plants and transferred to the atmosphere by the process of transpiration. These processes are collectively known as evapotranspiration.

Water that contacts the land and is not evaporated or taken up by plants either becomes runoff (flow across the ground surface) or infiltrates into the ground to become groundwater. The surface flow collects into drainages and progressively larger watercourses, ultimately forming rivers and lakes. Surface water that does not evaporate or infiltrate into the ground eventually reaches the ocean. Groundwater can flow back into the surface-water system through springs, seepage, or gaining streams and rivers (where the river elevation is below the aquifer elevation). Groundwater can be withdrawn through wells, reentering surface water for varying purposes

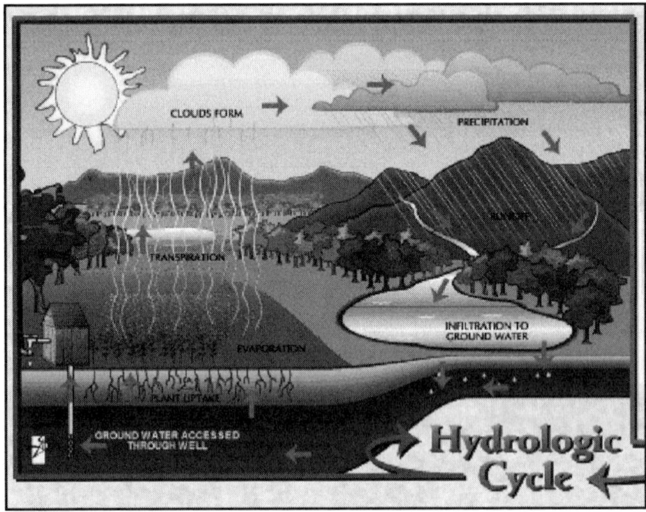

Source: Adapted from EPA 2003.

FIGURE 5.1 Hydrologic cycle

TABLE 5.1 Major inflow and outflow components of the hydrologic cycle

Inflow	Outflow
Precipitation	Evapotranspiration
Surface water flow	Surface water
Groundwater flow	Groundwater
Imported water	Exported water
Infiltration from irrigation, other activities	Consumptive use
Injection wells	Extraction wells

Source: Adapted from Weight and Sonderegger 2001.

(such as irrigation, industrial, and municipal uses). Although underground water interacts with geologic materials, including mineral deposits, groundwater may pick up acidity and metals through natural processes and can then reenter surface water as naturally impacted waters through springs or gaining streams, or as MIW if the mineral deposit has been mined.

SOLUTIONS AND MINERAL SOLUBILITY

Fundamental concepts of mineralogy and chemistry are useful in order to understand the effects that mining activities can have on drainage quality, although it is beyond the scope of this handbook to provide details on solution and mineral chemistry. Several more-detailed discussions of aqueous geochemistry fundamentals and geochemistry of MIW can be found in works by Garrels and Christ (1965), Evangelou (1998), Nordstrom (1999), Nordstrom and Alpers (1999), Plumlee et al. (1999), K.S. Smith (1999), and K.S. Smith and Huyck (1999).

Chemical reactions, including dissolution and precipitation, take place because of the chemical and electrochemical properties of elements, minerals, and biological activity. Chemical reactions can be expressed as chemical equations. Those referenced in this handbook are presented in the appendix.

Solubility is a measure of the amount of a material that can be dissolved in a solution (or *solvent*) of a given composition at a specified temperature and pressure (Drever 1988; Susak 1994; Williams-Jones et al. 1994). The solubility of a mineral or compound depends on numerous chemical properties of the mineral and on pressure, temperature, complex mineral–water interactions, water chemical composition, and secondary mineral formation. A solution is *saturated* when no more of a certain mineral will dissolve in the solution, no matter how much of the mineral is added. Solubility values generally identify the composition of the solvent (e.g., pure water, NaCl solution, sulfuric acid) as a specific temperature and pressure. For instance, the solubility of hydrated gypsum in cold water is 0.241 g $CaSO_4{\cdot}2H_2O$ per 100 cc H_2O and the solubility of anhydrous table salt, NaCl, in hot water is 39.12 g NaCl per 100 cc H_2O.

The rate at which the solubility of a solid is reached is a function of the kinetics of mineral dissolution and precipitation. Some solids dissolve quickly, whereas others take tens of years or longer to dissolve. Supersaturated solutions can result when rates of mineral precipitation are slow, resulting in higher solute concentrations than would be predicted for conditions of solubility equilibrium. Mineral solubility in water depends on numerous factors, especially the ionic species and concentration of all ions in the water, pH, acidity or alkalinity, and temperature. The solubilities of minerals for a particular set of chemical equilibrium conditions can be calculated from thermodynamic relationships using geochemical models.

Colloids are suspended particles in the solution, with a size range between several nanometers and several millimeters that can affect the solubility of other materials in MIW (Ranville and Schmiermund 1999). Adsorption of ions onto either colloids or minerals can also affect the solubilities of minerals that contain the same ions (Evangelou 1998).

ACIDITY AND ALKALINITY IN ROCKS

The tendency for natural solutions to dissolve minerals and produce drainage with elevated metals concentrations is measurable, in part, by pH, acidity, and alkalinity. The pH of a solution is the negative logarithm of the hydrogen ion (H^+) concentration and is expressed as presented in Equation 1 (see appendix), where values in brackets express molar concentration. The measurement of pH at a glass electrode, however, is related to the hydrogen ion activity, not concentration. Activity is not the same as concentration but is the effective concentration as defined by the chemical potential of a dissolved element. In dilute waters where interactions between ionic species are minimal, the activity of a dissolved element approximates its concentration, making measurement of pH at a glass electrode a useful tool. In concentrated solutions, such as seawater and brines, activity is a function of the charge of the dissolved element and the ionic strength of the solution. The ionic strength is related to the concentrations and ionic charges of all ionic species found in the fluid.

Although low pH is sometimes stated to be the same as acidity, this is not correct. Acidity is determined by measuring the concentration of hydroxide (OH^-) necessary to reach a specified pH under a specific set of titration conditions. The titration endpoint is usually at a pH of either 7.0 or 8.2, depending on the method, and measured acidity concentration is reported in milligrams per liter $CaCO_3$ (Drever 1988; Evangelou 1995; Schmiermund and Drozd 1997). During an acidity titration, hydroxide can be consumed by reactions with dissolved metals and cations as well as hydrogen ions, such that the acidity will generally exceed what would be calculated from only the pH. *Acidity*, therefore, is a measure of the hydroxide that can be consumed by hydrogen ions and by other cations, including metal ions, in the solution to reach a particular pH. Acidity is sometimes called mineral acidity or strong acidity (Drever 1988) when a relatively low pH endpoint is used in the acidity titration. In MIW, most mineral acidity comes from the cations of hydrogen (H^+), iron [Fe (II), Fe(III)], manganese [Mn(II)], and aluminum [Al(III)], though other metal cations can also contribute (appendix, Equations 2–6). When measuring acidity in MIW, it is important to thoroughly oxidize the solution using hydrogen peroxide before titrating, in order to measure the contributions from all of the metals to the acidity, in particular, ferrous ion [Fe(II)] and Mn(II). These metal cations consume hydroxide during hydrolysis and release hydrogen ions, producing acid. The contribution of aluminum species is called the aluminum acidity. Iron and manganese can substitute for aluminum in Equations 3–6 (see appendix).

The pH scale typically ranges from 0 to 14, but rare natural waters can have pH values outside of this range. MIW has been measured with a pH as low as –3.6 (Goleva et al. 1970; Nordstrom and Alpers 1999). Neutral pH (pH = 7.00) is the point where $[H^+] = [OH^-]$ and cannot be assumed to equal the background pH of natural waters. Most natural waters have a pH between 5 and 9 (Schmiermund and Drozd 1997). With a pH from 5 to 5.5, rainwater is naturally acidic because of the effects of atmosphere $CO_2(g)$ and naturally occurring acidic aerosols and particulates. Acid rain can have pH lower than 5 because of anthropogenic inputs of sulfur oxides and nitrogen oxides aerosols. Wetlands and saline lakes in arid climates can have pH greater than 9 because of evapoconcentration (Figure 5.2).

The general terms *acidic* and *basic* (or alkaline) waters refer to pH below (acidic) or above (basic) neutral pH (i.e., 7). However, water such as rainwater, in equilibrium with atmospheric CO_2, has a pH of 5.66 at 25°C (Drever 1988). Common usage of acidic drainage refers to waters with a pH below 6, because this is a water quality standard.

Strong *alkalinity* (e.g., sodium hydroxide) is the capacity of a solution to consume protons (hydrogen ions) or to neutralize an acid (appendix, Equations 7–9; Drever 1988; Schmiermund and Drozd 1997). The contribution of carbonate species to alkalinity is referred to as the carbonate alkalinity. In most natural waters, total alkalinity is equal to the carbonate alkalinity (appendix, Equation 9). Noncarbonate sources of alkalinity can include bisulfide, orthophosphate, ammonia, borate, salts of organic acids and silicic acid, and hydroxyl anions at pH greater than about 9 to 10. Commercial laboratories typically measure alkalinity by titrating the amount of acid

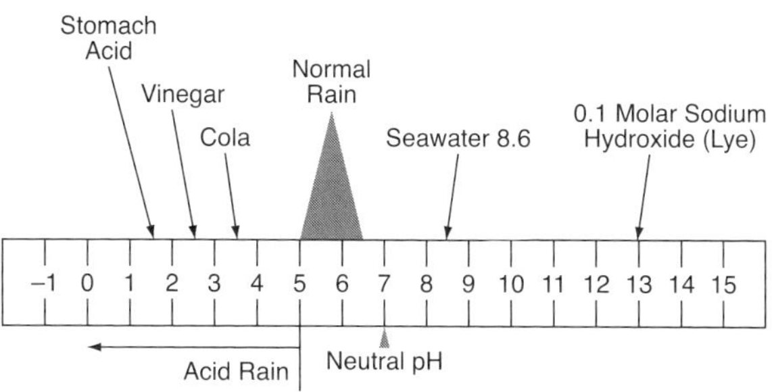

Courtesy of U.S. Geological Survey.

FIGURE 5.2 The pH scale showing various substances. Note unusual pH values <1 and >14.

required to bring a sample solution to a pH of 4.2 or to some other point specified by the client (Evangelou 1995).

The terms *acid-neutralizing capacity* and *base-neutralizing capacity* are equivalent to alkalinity and acidity, respectively, but as applied to solids, not solutions (Drever 1988). Terms, such as *acid neutralization potential* (ANP), *acid-generation potential* (AGP), *net neutralization potential*, and others are specifically defined for certain laboratory tests on solids rather than solutions. The AGP and ANP of rocks in contact with water and air are measured under controlled laboratory conditions or estimated from mineral phase analysis tests. Although a sample can have a balance of ANP and AGP, the rates of these acid generation and neutralization reactions in nature can be very different. Thus, the waters coming into contact with the solids can change significantly with time.

FORMATION OF ACID DRAINAGE

Acid drainage is a major water quality concern associated with metal mine wastes (mine waste rock piles, overburden piles, mill tailings, etc.) as well as with naturally occurring altered and mineralized rock. Mining typically increases the amount of oxygen in disturbed soils and rock, as well as increasing water mobility. By increasing the surface area of rocks as well as the pore volume around rocks, mining activity allows acid-producing reactions to occur more quickly. If the acid solution is not immediately neutralized by readily available alkalinity from the easily soluble carbonate minerals calcite and dolomite, then it can react with other minerals according to their physical availability and ease of dissolution. Acid drainage that exits mine waste disposal facilities can become a problem in surface drainages, soils, or groundwater.

Typically, acid drainage is high in iron and sulfate due to dissolution of iron sulfides and formation of sulfuric acid and iron ions. In some rocks, AD solutions contain high concentrations of aluminum, which is abundant in the earth's crust in nearly all major rock-forming minerals and, therefore, is present in most rocks. Even limestone and dolostone (dolomite), which are inherent acid neutralizers, tend to produce solutions containing aluminum, because both rocks contain a small percentage of clay, which contains aluminum. Aluminum concentrations in water can be enhanced if there is high fluorine in the waters, through the formation of stable aluminum fluoride complexes.

Generation of AD is a complex process that involves chemical, biological, and electrochemical reactions that vary with environmental conditions and composition of the water, mine wastes, ore, and host rocks.

TABLE 5.2 Important metal sulfides that occur in mining regions, in order of general abundance. The predominant acid producer is pyrite.

Mineral	Formula	Mineral	Formula
Pyrite	FeS_2	Molybdenite	MoS_2
Marcasite	FeS_2	Millerite	NiS
Pyrrhotite	Fe_xS_x	Galena	PbS
Chalcocite	Cu_2S	Sphalerite	ZnS
Covellite	CuS	Arsenopyrite	FeAsS
Chalcopyrite	$CuFeS_2$	Bornite	Cu_5FeS_4

The oxidation of pyrite is responsible for the majority of AD. The formation of AD generally involves five components: (1) air (oxygen), (2) iron (typically in pyrite, marcasite, pyrrhotite, or other sulfide or sulfate minerals; Tables 5.2 and 5.3; Figure 5.3), (3) sulfur (sulfide ions, sulfate ions, organic, native), (4) water, and (5) bacteria. These components are expressed by the following simplified equations:

iron sulfide + oxygen + water (+ bacteria) = sulfuric acid + ferric ions [Fe(III)]

iron sulfide + ferric ion + water = sulfuric acid + ferrous ions [Fe(II)]

These reactions are represented by Equations 10–14 (see appendix) defining sulfide oxidation (Nelson 1978; Stumm and Morgan 1981; Ritchie 1994; Evangelou 1995; Plumlee 1999). It is important to note that not much oxygen is needed if ferric ion [Fe(III)] is the primary oxidant for iron sulfides as it often is for systems with pH less than about 3.3 (Williamson et al. 2006).

Bacteria, such as *Acidithiobacillus ferrooxidans* and *Leptospirillea*, derive their metabolic energy from oxidizing ferrous ion [Fe(II)] to ferric ion [Fe(III)] and, in the process, utilize the sulfur and iron in their system to oxidize pyrite and produce AD. One of the most important reactions involving bacteria, typically in the pH range of 2.5 to 3.5, is the oxidation of Fe(II) to Fe(III) (appendix, Equation 15; Evangelou 1995; Mills 1998a).

In some cases, sulfide oxidation by the Fe(III) produced by bacteria can have reaction rates exceeding six to eight orders of magnitude greater than the same reactions without bacteria (Evangelou and Zhang 1995; Evangelou 1995). Thus, the bacteria accelerated AD production. Maintaining high pH (e.g., pH $>$ 3.5) can prevent bacterially mediated production of Fe(III) from accelerating sulfide oxidation, thus reducing the rate of acid production.

Fe(III) (appendix, Equations 10d and 16) is a powerful oxidizing agent for most metal sulfides (Mills 1998a, b). As long as Fe(III) is present, pyrite will continue to oxidize, even in the absence of oxygen (Evangelou 1995; Evangelou and Zhang 1995). The presence of manganese oxides at pH less than or equal to 5.5 can oxidize Fe(II) to Fe(III) and can form AD in some environments (appendix, Equations 16 and 17; Evangelou 1995). In some cases, AD also can form by reactions between sulfide minerals and Fe(III) without oxygen, such as in groundwaters (Rimstidt et al. 1994) (summarized by Equations 18–23, appendix).

Acidic drainage can also develop by the dissolution of secondary sulfate minerals (Table 5.3), which can form as intermediate oxidation products by precipitating from sulfate-rich solutions or by evaporation of mineralized solutions. In many deposits, secondary sulfate minerals may have formed naturally during acid alteration processes associated with sulfide mineralization. Soluble secondary sulfate minerals can be important as sinks of sulfuric acid, iron, trace metals, and sulfate in solid phases during dry periods. These sulfate minerals can be dissolved during wet periods, releasing the metals and forming sulfuric acid. This mechanism for the release of stored acidity due to sulfate mineral dissolution can result in the persistence of AD even in the absence

TABLE 5.3 Selected sulfate minerals

Selected Soluble Iron-Sulfate Minerals		Selected Soluble Sulfate Minerals		Less Soluble Sulfates (selected minerals of the alunite group)	
Mineral	**Formula**	**Mineral**	**Formula**	**Mineral**	**Formula**
Fe^{II}		Epsomite	$MgSO_4 \cdot 7H_2O$	Jarosite	$KFe_3^{III}(SO_4)_2(OH)_6$
Melanterite	$Fe^{II}SO_4 \cdot 7H_2O$	Hexahydrite	$MgSO_4 \cdot 6H_2O$	Natrojarosite	$NaFe_3^{III}(SO_4)_2(OH)_6$
Ferrohexa-hydrite	$Fe^{II}SO_4 \cdot 6H_2O$	Goslarite	$ZnSO_4 \cdot 7H_2O$	Hydronium jarosite	$(H_3O)Fe_3^{III}(SO_4)_2(OH)_6$
Siderotil	$Fe^{II}SO_4 \cdot 5H_2O$	Bianchite	$ZnSO_4 \cdot 6H_2O$	Ammoniojarosite	$(NH_4)Fe_3^{III}(SO_4)_2(OH)_6$
Rozenite	$Fe^{II}SO_4 \cdot 4H_2O$	Gunningite	$ZnSO_4 \cdot H_2O$	Argentojarosite	$AgFe_3^{III}(SO_4)_2(OH)_6$
Szomolnokite	$Fe^{II}SO_4 \cdot H_2O$	Gypsum	$CaSO_4 \cdot 2H_2O$	Plumbojarosite	$Pb_{0.5}Fe_3^{III}(SO_4)_2(OH)_6$
Halotrichite	$(Fe^{II})Al_2(SO_4)_4 \cdot 22H_2O$	Anhydrite	$CaSO_4$	Alunite	$KAl_3(SO_4)_2(OH)_6$
Mixed Fe^{II}-Fe^{III}		Retgersite	$NiSO_4 \cdot 6H_2O$	Natroalunite	$NaAl_3(SO_4)_2(OH)_6$
Copiapite	$Fe^{II}Fe_4^{III}(SO_4)_6(OH)_2 \cdot 20H_2O$	Chalcanthite	$CuSO_4 \cdot 5H_2O$	Ammonioalunite	$(NH_4)Al_3(SO_4)_2(OH)_6$
Bilinite	$Fe^{II}Fe_2^{III}(SO_4)_4 \cdot 22H_2O$	Alunogen	$Al_2(SO_4)_3 \cdot 17H_2O$	Osarizawaite	$PbCuAl_2(SO_4)_2(OH)_6$
Romerite	$Fe^{II}Fe_2^{III}(SO_4)_4 \cdot 14H_2O$	Mirabilite	$Na_2(SO_4) \cdot 10H_2O$	Beaver	$PbCuFe_2^{III}(SO_4)_2(OH)_6$
Voltaite	$K_2Fe^{II}_5Fe_4^{III}(SO_4)_{12} \cdot 18H_2O$	Thenardite	$Na_2(SO_4)$	—	—
Fe^{III}		—	—	—	—
Coquimbite	$Fe_2^{III}(SO_4)_3 \cdot 9H_2O$	—	—	—	—
Kornelite	$Fe_2^{III}(SO_4)_3 \cdot 7H_2O$	—	—	—	—
Rhomboclase	$HFe^{III}(SO_4)_2 \cdot 4H_2O$	—	—	—	—
Ferricopiapite	$Fe_5^{III}(SO_4)_6O(OH) \cdot 20H_2O$	—	—	—	—

Source: Alpers et al. 1994.

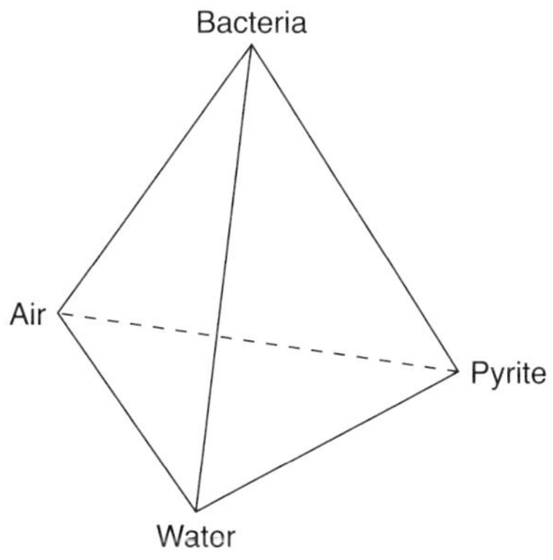

FIGURE 5.3 Acid drainage tetrahedron, showing the relationship between the four components that produce AD

of oxygen (Cravotta 1994). The dissolution of coquimbite ($Fe_2^{III}(SO_4)_3 \cdot 9H_2O$) (appendix, Equation 24) and halotrichite ($(Fe^{II})Al_2(SO_4)_4 \cdot 22H_2O$) (appendix, Equation 25) illustrates this effect. Hydrolysis of metal cations produced by sulfate mineral dissolution also produces acid (appendix, Equations 26 and 27).

Grain size and other textural variations affect the dissolution of secondary sulfate minerals. Some secondary sulfate minerals, such as coarse-grained anglesite ($PbSO_4$), are relatively insoluble and can store the metals and prevent their release to the environment (Alpers et al. 1994; Munroe et al. 2000). In contrast, fine-grained anglesite can be a source of dissolved lead, depending on solution conditions (K.S. Smith et al. 2000).

The rate of iron sulfide oxidation and attendant acid production depends on solid-phase compositional variables, microbial activity, the availability of oxygen and water, and texture (grain size, effective surface area, dispersion). Oxidation rates vary among sulfide minerals and are generally reported in decreasing order of reactivity as follows: marcasite > pyrrhotite > pyrite (e.g., Kwong and Ferguson 1990). However, different reactivity rankings have been reported by other authors and can be a function of mineral solubility, reaction conditions, trace element concentrations of the minerals, and crystal-morphology characteristics, among other factors (Jambor 1994; Plumlee 1999). For a given sulfide mineral, the oxidation rate increases with the increase in available surface area (i.e., decrease in grain size). For example, the oxidation of framboidal pyrite, with a high, associated surface area, is reported to be much more rapid than that of euhedral pyrite (Pugh et al. 1984; White and Jeffers 1994).

The rate of abiotic pyrite oxidation by oxygen decreases slightly as pH decreases. However, the overall abiotic rate increases as pH decreases into a range where ferric iron becomes the dominant oxidant (Williamson and Rimstidt 1994; Williamson et al. 2006). Nordstrom (1982) reported that as pH decreases to 4.5, ferric iron becomes more soluble and begins to act as an oxidizing agent. As pH further decreases, bacterial oxidation of ferrous to ferric iron becomes the rate-limiting step in the oxidation of pyrite (Singer and Stumm 1970), and ferric iron is the primary oxidizing agent for pyrite at pH below 4.5 (Nordstrom 1982; Singer and Stumm 1970; Kleinmann et al. 1981). In laboratory tests conducted by Lapakko and Antonson (1994), the oxidation rate of pyrrhotite in the pH range of 3.5 to 4.05 was roughly six to seven times greater than rates for the pH range of 5.35 to 6.1. The higher rate for the lower pH range was attributed to bacterially mediated oxidation of ferrous to ferric iron and consequent oxidation of the pyrrhotite by ferric iron. Data presented by Nordstrom and Alpers (1999) suggest that the bacterially mediated rate of pyrite oxidation by ferric iron is roughly two to three orders of magnitude faster than abiotic oxidation by oxygen at pH 2.

These weathering reactions produce acidic, iron, and sulfate-rich waters that can react with sulfide minerals and accelerate their oxidation; evaporate partially or totally to precipitate hydrated iron-sulfates and other minerals; and/or contact host rock minerals, which react to neutralize some or all of the acid. Acidic water flow, which is not neutralized within either the mine waste or host rock, will form acid drainage.

Hydrated iron-sulfate and trace-metal sulfate minerals (Table 5.3) precipitate during the evaporation of acidic, metal-rich, and sulfate-rich water within mine rock materials and store acid and metals released by sulfide mineral oxidation. The stored acid and metals can be subsequently released by additional flow through the mine rock (e.g., rain events, snow melt). The more common hydrated iron-sulfate minerals that occur as efflorescent salts due to the weathering of pyrite include melanterite ($Fe^{II}SO_4 \cdot 7H_2O$), rozenite ($Fe^{II}SO_4 \cdot 4H_2O$), szomolnokite ($Fe^{II}SO_4 \cdot H_2O$), romerite ($Fe^{II}Fe^{III}_2(SO_4)_4 \cdot 14H_2O$), and copiapite ($Fe^{II}Fe_4^{III}(SO_4)_6(OH)_2 \cdot 20H_2O$), respectively (Alpers et al. 1994). According to Nordstrom

(1982) and Cravotta (1994), these efflorescent salts are highly soluble and provide an instantaneous source of acidic water upon dissolution and hydrolysis. They are partially responsible for increased acidity and metals loadings in the receiving environment during rain events and periods of high rates of snow melting. Their cumulative storage and incremental release can help explain the lag from mine waste placement to AD formation, particularly in arid climates.

In most rock–water systems, the primary acid-consuming minerals are the carbonates, including calcite ($CaCO_3$) and dolomite [$CaMg(CO_3)_2$] (appendix, Equations 28–31), and less commonly magnesite ($MgCO_3$). Reactions producing bicarbonate (appendix, Equations 28 and 30) are the predominant ones describing acid neutralization by reaction with carbonates for pH > 6.4 for systems open to the atmosphere. Reactions producing carbonic acid (appendix, Equations 29 and 31) are predominant below pH 6.4 (Drever 1988; Lapakko et al. 1998).

Of the carbonate minerals, calcite dissolves most rapidly (Busenberg and Plummer 1986). Relative to calcite, the rate of dolomite dissolution is about an order of magnitude slower (Busenberg and Plummer 1982), and the rate of $MgCO_3$ dissolution is about four orders of magnitude slower (Chou et al. 1989). The rate of siderite ($FeCO_3$) dissolution under anoxic conditions is reported to be three orders of magnitude slower than that of calcite (Greenberg and Tomson 1992). However, iron and manganese carbonates (rhodochrosite, $MnCO_3$) do not provide net acid neutralization due to reactions that may occur later under oxidizing conditions. Oxidation of the released iron or manganese and the subsequent acid production by hydrolysis and precipitation reactions, as hydroxides, result in the precipitation of the hydroxide from solution.

Some reactions of silicate minerals will neutralize acid, such as dissolution of potassium feldspar (appendix, Equation 32), plagioclase (Busenberg and Clemency 1976), epidote, olivine (Hem 1970), or muscovite. However, their rates of dissolution and consequent acid neutralization are very slow relative to the carbonate minerals (appendix, Equation 33; Ritchie 1994; Nesbitt and Jambor 1998; White et al. 1999). For example, White et al. (1999) noted that "at near neutral pH, the dissolution rate of calcite is approximately seven orders of magnitude faster than the dissolution of plagioclase feldspar." The effectiveness of silicate minerals in neutralizing acid increases with increasing silicate mineral content and decreases with grain size (Morin and Hutt 1994). Also, the silicates contained in mafic rocks (e.g., diorites, gabbros, peridotites) generally react much more rapidly than the silicates contained in silicic rocks (e.g., granites and monzonites) and can delay or prevent AD under some circumstances (Eary and Williamson 2006; Jambor, 2002; Lapakko et al. 1997). For example, silicate minerals have been reported to be quite effective in neutralizing acid rock drainage (i.e., Boulder River, Montana; Desborough et al. 1998). Dissolution of silicate minerals can also result in the formation of less reactive solid phases, such as the formation of kaolinite from plagioclase (appendix, Equation 34) and illite from potassium feldspar. Adsorption of hydrogen ion onto mineral surfaces releases adsorbed or ion-exchanged cations and also consumes acid.

Trace metals occur at low average concentrations in the earth's crust but can be present at elevated levels in mineralized areas. They are commonly found as sulfide minerals, the oxidation of which can release the trace metal from the relatively insoluble sulfide phases. Once released to solution, several types of reactions can influence the migration and fate of these metals. K.S. Smith and Huyck (1999) present a series of diagrams for the generalized relative mobility of elements under different environmental conditions for use as an initial estimate of metal behavior in surficial environments. At a regional scale, generalizations can frequently be used to estimate broad trends in metal mobility. However, as the scale becomes increasingly smaller, estimating metal behavior generally becomes increasingly difficult (K.S. Smith and Huyck 1999).

In general, metals either remain in solution or are removed in secondary phases. For removal from solution, trace metals can precipitate as oxides, hydroxides, or carbonates. They can be adsorbed by surfaces such as iron oxyhydroxides (K.S. Smith 1999) or coprecipitate with other solid phases. In acidic solutions, trace metal removal is limited, and elevated trace metal concentrations are often related to acidity. However, circumneutral drainages (pH 5–9) also can contain elevated concentrations of trace metals such as nickel, copper, cobalt (Lapakko 1993), zinc, manganese (K.S. Smith and Huyck 1999), molybdenum (Brown 1989), arsenic, selenium, and antimony. Concentrations of molybdenum, arsenic, selenium, and antimony, in particular, can be elevated in MIW even as pH increases above 7, due to the stability of the oxyanions at high pH.

Oxidation of arsenic, selenium, and antimony sulfides can produce weak acids, and sulfuric acid can be produced from oxidation of the iron sulfide fraction of mixed sulfide minerals such as chalcopyrite (Plumlee 1999). Oxidation of most other trace metal sulfides can produce acid only if the metal released hydrolyzes or precipitates as a hydroxide, oxide, or carbonate. For most trace metals this will occur only at pH levels above 6, and, as pH decreases below this level, the secondary phases will dissolve. Consequently, these phases generally do not contribute to acid production observed at lower pH levels.

Whereas the acid-producing and acid-neutralizing mineral contents influence mine rock and metallurgical material drainage quality, several subtle mineralogical factors also are influential. Individual minerals can be entirely liberated from the rock matrix or occur interstitial to other minerals (partially liberated) or as inclusions within other minerals. The extent of liberation affects availability for reaction. For example, acid-producing or acid-neutralizing minerals encapsulated within minerals such as quartz will essentially be unavailable for reaction.

Oxidation of sulfide minerals and dissolution of carbonate minerals are surface reactions and, therefore, the rates of these reactions are dependent on the reactive surface area. Reactivity decreases as mineral surfaces are covered with coatings, such as iron oxyhydroxides, whereas the concentration of lattice defects tends to increase reactivity. Mineral surface area is dependent on the extent to which the mineral is liberated from the rock matrix, mineral grain size, and the roughness of the mineral surface.

FORMS OF SULFUR IN ROCKS

Sulfur in rocks associated with metal mines is found as organic sulfur, sulfate minerals, sulfide minerals, and native sulfur. Organic sulfur is believed to be complexed and combined with organic compounds in carbonaceous rocks, which are not present in many metal ore deposits. Generally, the organic sulfur component is not chemically reactive and has little or no direct effect on acid-producing potential.

Sulfate sulfur from oxidation of sulfides is typically found in variable quantities in freshly exposed sulfide ores and wastes and other pyrite-containing rocks. Some sulfate minerals like jarosite $[KFe_3(SO_4)_2(OH)_6]$ can dissolve and form acid solutions in near-surface environments at low pH (Ford 1929). Sulfates such as barite and alunite are quite common in many gold ore deposits.

Pyrite, the predominant sulfide present in metallic ores and associated wastes, can vary in those rocks from very small (<10 μm) crystalline grains disseminated in the rock matrix to large (>2 mm), well-formed crystals or masses (e.g., veins) easily visible to the naked eye. Most of the pyrite in metalliferous ore zones is associated with hydrothermal activity that is typically related to ore deposition.

Pyrite is the sulfur form of greatest concern for AD. Accordingly, the potential acidity of a fresh rock sample is estimated by the pyritic sulfur content, not the total sulfur content. Studies have shown that variations in total sulfur contents of some mine waste samples reflect similar variations in pyritic sulfur content. Several types of pyritic sulfur are known based on physical appearance, and they are classed into six groups: (1) primary massive, (2) plant replacement pyrite, (3) primary euhedral pyrite, (4) secondary cleat (joint) coatings, (5) mossy pitted pyrite, and (6) framboidal pyrite. Caruccio and Geidel (1980) and Caruccio et al. (1988) provide an extensive review of the different forms, morphologies, and reactivity of pyritic materials.

The equations for pyrite oxidation show that a material containing 1% sulfur, as pyrite, would yield upon complete reaction an amount of sulfuric acid that requires 31.25 mg of $CaCO_3$ to neutralize 1 g (3.13% or 31.25/1,000 tons) of the material (Sobek et al. 1978). If pyrite oxidation involves ferric ion [Fe(III)], then more acid is produced, requiring proportionate $CaCO_3$ to neutralize the acid. When sulfur in the rock is exclusively pyrite, the total sulfur content of the rock accurately quantifies the acid-producing potential, if it were all to react. When either organic or sulfate sulfur are present in significant amounts in partially weathered rocks, total sulfur measurements overestimate the amount of acid that will be formed upon oxidation and should not be used. Therefore, correction for sulfates and organic sulfur naturally present in some mine wastes or resulting from partial weathering of pyritic materials is necessary to increase accuracy in predicting the acid-producing potential of materials containing mixed sulfur species.

The rate of pyrite oxidation depends on numerous variables:

- Reactive surface area of pyrite (Singer and Stumm 1970);
- Form of pyritic sulfur (Caruccio et al. 1988);
- Oxygen concentrations (McKibben and Barnes 1986; E.E. Smith and Shumate 1970);
- Solution pH (E.E. Smith and Shumate 1970; Williamson and Rimstidt 1994);
- Ferric iron concentration (McKibben and Barnes 1986; Williamson and Rimstidt 1994);
- Humidity (Fennemore et al. 1998; Jerz and Rimstidt 2004);
- Catalytic agents, including sorption by clay minerals (Caruccio et al. 1988);
- Water-flushing frequencies (moving of water through waste rock piles; Caruccio et al. 1988); and
- Presence of *Acidithiobacillus* and *Leptospirillea* bacteria (EPA 1971).

The possibility of identifying and quantifying the effects of these and other controlling factors with all the various rock types in a field setting is unlikely. A variety of approaches to evaluate the rate of oxidation is discussed in the other handbooks in the series.

Native sulfur in carbonate rocks associated with salt domes and in metal deposits sometimes occurs in quantities large enough to contribute to AD.

SOURCES OF ACID DRAINAGE

Acid drainage and other MIW can evolve from natural or human-made sources, or from human-caused disturbances (i.e., anthropogenic; Figure 5.4). AD results when acid production exceeds acid-consumption reactions, which are part of weathering or oxidation of natural materials exposed to water and air. Relative rates of acid-producing and acid-generating reactions are controlled in part by accessibility of minerals that contribute to these reactions.

Natural AD occurs in some specific geologic environments (see, for example, Plumlee 1999; Posey et al. 2000). Certain types of hydrothermal alteration, which accompany some types of hydrothermal ore deposits, have characteristic high-pyrite and low-carbonate concentrations that together foster AD. Recognizing these features prior to mining can provide many indications about anticipated water quality and local groundwater hydrology. Weathering and erosion of

FIGURE 5.4 Oxidized waste piles can be sources of MIW (Comet mine site, Boulder River, Montana)

pyrite-bearing, hydrothermally altered rocks can produce badlands topography, termed *alteration scars*, partially as a result of natural AD and associated erosion (Meyer and Leonardson 1990). Waste rock piles from open pit mining in the foreground of Figure 5.5 overlie some scar areas. These scars include exposures where hydrothermal alteration, weathering, and rapid erosion preclude vegetation growth due to naturally acidic soils and AD-related surface water and seeps where iron hydroxide and other hydroxide precipitates form.

Some sedimentary and volcanic rock units also have characteristically high iron sulfide (pyrite, pyrrhotite, marcasite) concentrations that lead to local AD where they are exposed to weathering. Many black shale deposits that form in reducing, oxygen-poor environments tend to contain or are associated with adjacent layers of rock that are high in iron sulfides. Where such sulfide occurrences are exposed to elevated rates of physical weathering, and in the absence of calcite or dolomite, such deposits can generate local AD. Where natural AD is associated with mining or mineral-processing operations, it can still be considered to be MIW, but the impact assessment of those waters must take into consideration the natural conditions prevailing at the site prior to operations.

Mines are not the only source of AD. Physical disturbances of rock such as highway road cuts, subway and highway tunnels, logging, or other construction activities can create AD as long as the rock contains sulfide minerals, particularly pyrite, that are not readily neutralized upon weathering and formation of AD.

METAL TRANSPORT WITH NEUTRAL AND ALKALINE pH

Not all MIW is acidic. Alkaline or neutral drainage that is high in metals commonly develops when metal-rich, sulfuric acid solutions are neutralized by reaction with either calcite or dolomite. The concentration of any element in drainage derived from weathering of sulfide minerals is a function of the concentration of the elements in ore or host rock minerals, exposure of surface area of the minerals (porosity, grain size, fracture density, broken surfaces), susceptibility of

Courtesy of Molycorp, Inc.

FIGURE 5.5 View from the air looking eastward at spectacular examples of natural alteration scars (left background) proximal to Molycorp's Questa mine north of Red River, New Mexico, in the middle ground

the mineral to weathering and oxidation (climate combined with mineral properties), and mobility of the element under surface or near-surface chemical conditions (solution and mineral solubility chemistry).

Neutral and alkaline waters, which also are associated with mining (i.e., MIW), can react with and dissolve minerals and transport metals. Amphoteric metal hydroxides, which include but are not limited to aluminum, antimony, arsenic, beryllium, chromium, fluorine, selenium, thallium, tin, and zinc, and to a lesser extent copper and silver (Mortimer 1967), dissolve in both high- and low-pH waters. Increases in water temperatures typically increase the solubility of metals, thereby allowing neutral to slightly alkaline waters to mobilize metals. Some metals (e.g., uranium), industrial minerals (e.g., phosphate), and other mineral-processing reagents (e.g., cyanide) are more soluble in neutral to alkaline waters. One important mechanism in metal mobility is the presence and composition of colloids or suspended matter. Brines, highly concentrated solutions composed mostly of salts, are also capable of mobilizing metals and affecting drainage quality, along with elevated salt concentrations.

MINING AND METALLURGICAL PROCESSES AND CHEMICAL REAGENTS

The use of explosives and cyanide are examples of mining and metallurgical processes that have the potential to adversely affect drainage quality if the materials or reagents used for each process are not properly managed.

Blasting Materials

Nitrate (NO_3^-) represents the most oxidized chemical form of nitrogen found in natural systems A negatively charged anion, it must be paired with a positively charged cation, as in the salts potassium nitrate ($K^+NO_3^-$) or sodium nitrate ($Na^+NO_3^-$), or other combinations. Nitrate is one of the most water-soluble anions known.

Ammonium nitrate is primarily used as a fertilizer but also is widely used with additives as a blasting agent in the mining industry: for excavation of shafts, tunnels, and galleries; for mining of nonmetallic material such as limestone, ironstone, and road metal; and for other purposes. Because of the dissolution of nitrates from the residue of the explosives, such usage can lead to environmental impacts on both surface and groundwater. In water, the accumulation of nitrate can impact human health and aquatic life. Even a relatively low level is a potential health threat to infants. The current EPA maximum contaminant level (MCL) for nitrate as nitrogen is 10 mg/L (EPA 2006). In the environment, excess nitrate in open water can cause algae blooms, which create a shallow mat that absorbs the sunlight required for photosynthesis carried out by the deeper aquatic plants, and an excess amount of organic matter that consumes oxygen when it decays. Many organisms, such as clams and scallops (macroinvertabrates), that are dependent on oxygen respiration may die because of oxygen depletion related to nitrate-induced algal blooms, since no oxygen is released by the deeper water plants by photosynthesis. Therefore, it is important to identify the types of blasting agents used and where the components of the blasting media may result in adverse effects on drainage quality.

Cyanide

Cyanide is both a naturally occurring compound produced biochemically and a synthetic chemical. *Cyanides* is sometimes used as a general term for chemical compounds containing cyanide ion (CN^-), hydrogen cyanide gas (HCN), and/or metal cyanide complexes. In nature, cyanide occurs in many common plants, including lettuce, broccoli, sweet potatoes, lima beans, almonds, and tobacco (A. Smith and Struhsacker 1988). In some plants (cassava, for instance, which is used to make tapioca), cyanide occurs in natural concentrations that are toxic and thereby requires elaborate preparation to remove cyanides to safe levels (MEM 2001). HCN is produced commercially for use in production of adhesives, computer electronics, fire retardants, cosmetics, dyes, nylon, paints, pharmaceuticals, acrylic plastic sheets, rocket propellants, and road and table salts (Mudder 2001).

Approximately 1.4 million tons of hydrogen cyanide gas are produced annually, worldwide, of which approximately 13% is converted into either sodium cyanide (NaCN) or calcium cyanide ($Ca(CN)_2$), which are used in the extraction of gold, silver, and other metals and in the plating of metal coatings. The metallurgical use of sodium cyanide in cyanide leaching has revolutionized the recovery of gold and other metals. Cyanides are also used in other metallurgical applications, typically in froth flotation circuits where they are used to depress pyrite and to segregate certain sulfide minerals from others.

In 1887, scientists J.S. MacArthur, R.W. Forrest, and W. Forrest, in Glasgow, Scotland, patented the cyanide leaching process (Dorr and Bosqui 1950; Marsden and House 2006). The first commercial cyanide plant was at the Crown gold mine in New Zealand in 1889. The first plants in the United States appeared in 1891 at Mercur, Utah, and Calumet, California (Dorr and Bosqui 1950). Between 1892 and 1905, U.S. gold production increased from 1.7 million oz to 4.6 million oz due to retreatment via cyanidation of tailings previously stripped of gold via gravity separation and mercury amalgamation. Today, cyanide leaching is the most commonly used method for gold extraction and is applied at dump and heap leaching operations and at mills in stirred tanks and vats.

In conventional cyanide heap leaching, the ore is crushed and piled onto a lined leach pad to which a weak (50–500 ppm) cyanide solution is applied to the ore surface, either through tricklers or sprayers. The most common cyanide compound used in leaching is sodium cyanide, but calcium cyanide and potassium cyanide (KCN) are also used. As the cyanide solution seeps downward through the ore, it reacts with gold and other metals, such as silver, mercury, copper, and nickel, and forms a solution of metal cyanide complexes. The resulting "pregnant solution," containing metal cyanide complexes that have leached through the ore, is captured on the liner by gravity and treated to remove the gold and, perhaps, other metals (appendix, Equations 35 and 36; A. Smith 1994; EPA 1994; A.C.S. Smith and Mudder 1999). After gold recovery, the "barren solution" usually contains unreacted cyanide, reduced concentrations of metal–cyanide complexes, and cyanide degradation products such as cyanate, thiocyanate, ammonia, and nitrates. This solution is regenerated with more cyanide and alkaline additives, usually lime, and then recycled back to the heap for further leaching. When environmental conditions at a site (e.g., excess rainfall on the heap leach facility) result in an excess of water, the barren solution may need to be detoxified to destroy the cyanide and remove the metals that were complexed with the cyanide and other anions (e.g., nitrates and oxyanions of arsenic, antimony, and selenium) before it can be released to the environment.

The amenability of ore to cyanidation depends on several mineralogical, metallurgical, and physical characteristics of the ore; these are discussed in some detail by Dorr and Bosqui (1950), Cole and Kirkpatrick (1983), Marsden and House (2006), A.C.S. Smith and Mudder (1999), and others.

At closure, the cyanide solutions and the waste piles must be detoxified or naturally degraded by extended rinsing of the spent leached material. Even though hydrolysis of cyanides in groundwater is slow, there are rare documented cases of cyanide in groundwater near some mining facilities (A.C.S. Smith and Mudder 1999). Most mine tailings and heap leach facilities show no impact of cyanide in routine groundwater monitoring (A.C.S. Smith and Mudder 1999).

Stirred tank cyanidation is chemically similar to heap leaching but involves leaching of finely crushed ores in an agitated, oxygenated tank. The finely ground ore is leached with cyanide to strip the target metals, and the pregnant solution is collected and treated or contacted with activated carbon to remove gold and silver. The leached slurry is discharged to the tailings ponds. In order to prevent impacts on wildlife, tailings slurries from leaching may require detoxification of the cyanide prior to discharge to the storage facility. Excess solution from the tailings are generally recycled to the milling process, but also can require detoxification and discharge to the environment when excess water causes an imbalance in the recirculating system.

Gold recovery from heap leaching or stirred tank leaching can be accomplished through direct contact of activated carbon with the slurries in tanks or solutions in columns. After the carbon is loaded with metal cyanide complexes by adsorption, it is separated from the slurry through screening in tanks or by pumping from the columns, and these complexes are stripped by desorption from the carbon with an alkaline cyanide and/or caustic solution. Whether carbon is loaded while in the tank or separately from contact with the pregnant solution, the carbon-stripping process generates a barren cyanide solution, which is recycled. Alternatively, the leached stirred tank solids can be separated from the solution, and the silver and gold recovered by zinc precipitation instead of by carbon adsorption.

The acute toxicity of cyanide in small doses is a significant industrial hygiene issue for cyanide use and requires careful adherence to safe handling procedures in order to prevent accidental exposure of personnel. The lethal dose for humans varies with the type of exposure (ingested vs. inhaled or absorbed) and type of cyanide compound (Mason 1998; Boening and Chew 1999). The EPA's MCL for free cyanide in drinking water is 0.2 mg/L (EPA 2006).

Mercury Amalgamation

Amalgamation is a chemical reaction that dissolves native gold and silver from the ores as it forms a gold and silver amalgam with mercury. To remove gold and silver from the amalgam, mercury is then heated and driven off, with the gold and silver remaining. The patio process, developed in Mexico in the 16th century, used mercury amalgamation to free silver from the ores and was largely replaced by the cyanide process in the 20th century.

Before the 1900s, mercury amalgamation was used almost exclusively to recover native gold and silver from oxide ores. The process is still used in the United States and other developed countries by hobby miners but not by the commercial metallurgical processing industry. Mercury amalgamation is commonly used by artisanal miners in underdeveloped countries where gold is present in visible quantities (Whitehouse et al. 2006). Although less effective than other processes at removing gold, mercury amalgamation's attraction is that it provides a simple, easily transportable means of recovering gold. It is used to recover fine free gold and silver in high-grade ores where the mercury can contact the precious metals directly, but it is not effective at recovering gold and silver from low-grade deposits or deposits in which the metals are locked within grains of other minerals.

Mercury is toxic at very low concentrations and persists in the environment, where it bioaccumulates in animal tissues, especially fish, which can be dangerous to consume. Mercury poisoning can result in death for humans and other biological organisms, and certain industrial and biological processes can concentrate mercury compounds to dangerous levels. For these reasons, it is not generally advisable to use mercury amalgamation unless strict recovery and disposal measures are employed.

CHAPTER 6

Potential Impacts of MIW on the Environment

If not properly managed, water supplies can be affected by drainage from mining and metallurgical facilities. Highly acidic and some alkaline waters can be unhealthy to organisms, especially fish, which have an optimum water pH range between 6.5 and 9 (Figure 6.1). Acidic and alkaline waters can also dissolve metals previously bound up in ore minerals and surrounding rock. Although some trace metals are essential for biological life, in extremely high concentrations they can be toxic. When metals enter aquatic environments, they can have devastating effects on aquatic and plant life, including birds, fish, and other aquatic organisms. Humans also can be affected by direct ingestion of MIW and by direct contact through outdoor activities such as swimming.

The potential effects of mining and metallurgical drainage on the environment can be broadly divided into six interrelated and complex categories of impacts:

1. Land surface,
2. Biological,
3. Hydrological,
4. Air quality,
5. Societal, and
6. Others from mining and metallurgical processing, such as additions of dangerous substances during blasting and processing (Allgaier 1997).

All of these environmental impacts can affect water and/or drainage quality, either directly or indirectly. The effects of mining vary tremendously among mine sites. Some can have only minimal effects, whereas other mine sites can seriously impact the water and drainage quality. Although geology and climate are among the main controls of MIW, the various phases of mining and cessation of mining also affect the drainage quality.

LAND SURFACE IMPACTS

Mining and metallurgical processing are necessary and temporary uses of land that require disruption and disturbance of the land surface and sometimes of the subsurface. The topography, ore deposit shape, mineral economics, climate, and type of mineral deposit play important roles in determining mining methods, extent of disturbance of the land surface, and impacts on water and drainage quality. Specific mining impacts that can degrade the water and drainage quality are erosion, sedimentation, subsidence, differential settling of landfills and regraded mine areas, exposure and possible releases of metals, and reshaping of geomorphic features.

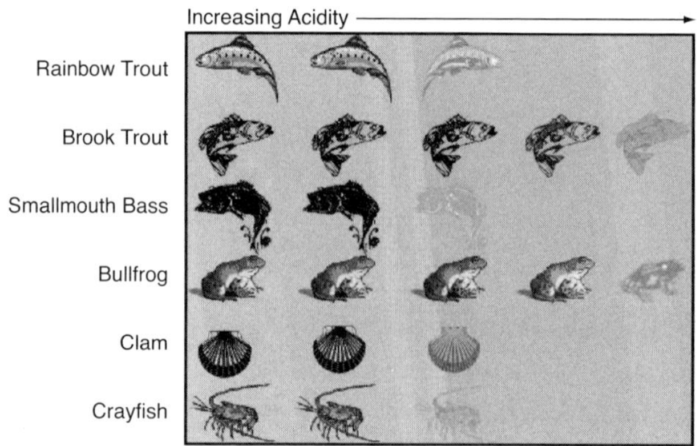

Courtesy of U.S. Geological Survey.

FIGURE 6.1 Effect of acidity on selected aquatic species

BIOLOGICAL IMPACTS

Mining and metallurgical processing's effects on vegetation are particularly important, because vegetation stabilizes the soil. Typically, mining requires the removal of vegetation and this can result in increased erosion and changes in visual aesthetics. Increased erosion; addition of heavy metals; loss of wetlands; and changes, fragmentation, or destruction of ecosystems or habitats can change the amount and type of vegetation cover, vegetation density, productivity, type of vegetation, diversity, and plant morphology.

Mining can impact wildlife in a variety of additional ways, including physical injury or mortality of species, changes in breeding and migration patterns, and disruption of food and water supplies. Endangered species, especially, are often vulnerable to mining activities. Another impact of mining is wetlands, which can be developed during mining operations or after mining ceases or can dry up during or after mining.

HYDROLOGIC IMPACTS

There are four major types of potential impacts from mining and metallurgical processing on water quality: (1) acid drainage (Figure 6.2), (2) metal leaching and resultant contamination, (3) release of processing chemicals, and (4) increased erosion and sedimentation.

Both surface and groundwaters can be affected. Pit lakes from open pit mining (Figure 6.3) have special considerations. These will be addressed in Volume 3 of this series, *Mine Pit Lakes: Characteristics, Predictive Modeling, and Sustainability.*

AIR QUALITY IMPACTS

All aspects of mining, from exploration, extraction, and beneficiation to reclamation, can have some measurable effect on the air quality. Emissions from various stages of mining and mineral processing can result in particulate matter, including gases, dust, and other substances, being released into the atmosphere that can create problems with the air we breathe. These particulates can move long distances by wind before being washed out of the atmosphere by precipitation or simple settling out of the atmosphere by gravity or absorption by plants or animals. Metallurgical processing can affect air quality due to emissions from smelters, sprayers, crushers, ore roasters,

FIGURE 6.2 Acid drainage–affected waters and waste rock, Leadville, Colorado

FIGURE 6.3 Pit lake at the Copper Flat mine, Hillsboro district, Sierra County, New Mexico. Many porphyry copper deposits are mined by open pit methods and result in pit lakes that have low pH values. The pH of this lake varies seasonally from 4 to 6.

autoclaves, refining operations, and tailings facilities. Before the modern use of scrubbers and stacks, which remove sulfur and particulate matter from the emissions, acid rain in some areas was attributed to smelter emissions.

SOCIETAL IMPACTS

Some of the most important issues that the general public associates with mining include direct hazards such as the environmental impacts to surface water and groundwater, generation of dust, smelter emissions, changes in slope stabilities around excavations, potentially dangerous working conditions, land subsidence, and undesirable visual impacts.

Perceptions by the public and public participation in the mine permitting and reclamation processes can dramatically affect how mining and metallurgical operations plan and implement reclamation efforts, pollution control systems, operational safeguards, and closure efforts. In addition, cultural resources, such as archaeological sites and recreation sites, must be protected without damaging water and drainage quality. The public must be involved at the beginning of any permitting process and throughout the life of the mining, processing, and reclamation of the site (discussed in more detail in Volume 6, *Sampling and Monitoring for the Mine Life Cycle*).

OTHER IMPACTS

Alteration of the land by blasting (i.e., flyrock, ground vibrations, airblast, air and rock pressure pulse) and subsidence created by removing material during mining can adversely alter the air, soil, water, and drainage quality in some local areas.

Erosion and Sedimentation

One of the more common legacies of mining, erosion is caused by physical and chemical processes, usually together, though locally weathering is generally dominated by one process or the other. Physical weathering includes undercutting outcrops, ice wedging, slope instability resulting in slope movements, and wind erosion. Chemical weathering dominates the weathering process where readily soluble minerals dissolve out of a matrix, leaving a weaker network of more chemically resistant yet physically unstable minerals. Weathering of silicate minerals can produce clay minerals, which reduces the shear strength of the rocks and can promote rockslides. Sedimentation problems can arise if eroded materials dam up drainage ways, alter drainage patterns, stress vegetation via overbank deposition, or degrade wetlands. Sediments can also affect aquatic habitats by coating eggs, submerging invertebrate habitat, and stressing aquatic vegetation.

Stability Issues

Surface subsidence is a natural consequence of some mining and occurs when strata overlying underground workings collapse into mined-out voids, typically as sinkholes or troughs. Sinkholes or depressions can interrupt surface water drainage patterns and affect ponds, streams, and wetlands. The impact and extent of subsidence is related to the method and depth of mining. Traditional room-and-pillar methods generally leave enough material in place to avoid subsidence, although shallow workings in weak rock units are still susceptible to subsidence. Typically, high-volume extraction techniques, such as pillar retreat, block and sublevel caving, and longwall mining, result in some amount of subsidence. Remedial measures include leaving support mechanisms (i.e., pillars) in place after completing the mining operation or backfilling with either cemented tailings or waste rock. Reducing the withdrawal of groundwater in some cases can reduce the potential for subsidence.

Dust Emissions

Mining operations can emit airborne particles, usually as fugitive dust. Sources of fugitive dust include crushing, conveyance of crushed material, loading bins, blasting, mine and motor vehicle traffic, haul roads, waste rock piles, windblown tailings, and disturbed areas in general. Depending on the mineralogy of the sources, dust can in some cases contain arsenic, lead, or other metals. Less commonly, asbestos can appear in fugitive dust. Particulate matter is an environmental concern, because it can contaminate air and deposit contaminated dust in surface water, causing sedimentation, turbidity, and, in extreme cases, affecting drainage quality. Weathered sulfide tailings can be a source of acid where soluble iron sulfate salts blow from the tails onto soils or water bodies downwind.

Other Chemicals of Concern

Other chemicals not directly related to mining and processing can be found at mine sites and could affect drainage quality. These can include petroleum products used for fuels, degreasers, and lubricants, laboratory wastes including lead from fire assay for gold and silver, septic and refuse systems associated with occupational work areas, construction materials, and soaps and detergents.

Appendix

Chemical Equations

$$pH = -\log [H^+] \quad \text{(EQ 1)}$$

$$OH^- + H^+ = H_2O \text{ or } OH^- + H_3O^+ = 2H_2O \quad \text{(EQ 2)}$$

Hydrolysis reactions, especially aluminum and iron, contribute to the acidity:

$$Al^{3+} + 3H_2O = Al(OH)_{3(s)} + 3H^+ \text{ and } Fe^{3+} + 3H_2O = Fe(OH)_{3(s)} + 3H^+ \quad \text{(EQ 3)}$$

$$Al^{3+} + 2H_2O = Al(OH)_2^{\ +} + 2H^+ \quad \text{(EQ 4)}$$

$$Al^{3+} + H_2O = Al(OH)^{2+} + H^+ \quad \text{(EQ 5)}$$

and the acidity becomes (Drever, 1988):

$$\text{Acidity} = m[H^+] - m[HCO_3^{\ -}] + 3m[Al^{3+}] + 2m[Al(OH)^{2+}] + m[Al(OH)_2^{\ +}] + m[Al(OH)_4^{\ -}] \text{ (plus any Fe and Mn component)} \quad \text{(EQ 6)}$$

Alkalinity is the capacity of a solution to consume protons (hydrogen ions) or to neutralize an acid and is expressed as (Drever 1988; Schmiermund and Drozd 1997):

$$CO^-_3 + H^+ = HCO^-_3 \quad \text{(EQ 7)}$$

$$HCO^-_3 + H^+ = H_2CO_3 \quad \text{(EQ 8)}$$

And the alkalinity becomes (Drever 1988):

$$\text{Alkalinity} = m[HCO_3^{\ -}] + 2m[CO_3^{\ 2-}] \quad \text{(EQ 9)}$$

Alkalinity in mining influenced water can involve more than carbonate (e.g., $H_3SiO_4^{\ -}$).

Formation of acid drainage:

$$FeS_{2(s)} + 3.5O_2 + H_2O = Fe^{2+} + 2SO_4^{\ 2-} + 2H^+ \quad \text{(EQ 10A)}$$

$$FeS_{2(s)} + 3.75O_2 + 0.5H_2O = Fe^{3+} + 2SO_4^{\ 2-} + H^+ \quad \text{(EQ 10B)}$$

$$FeS_{2(s)} + 3.75O_2 + 3.5H_2O = Fe(OH)_{3(s)} + 2SO_4^{\ 2-} + 4H^+ \text{ at pH} = 4 \quad \text{(EQ 10C)}$$

$$FeS_{2(s)} + 14Fe^{3+} + 8H_2O = 15\ Fe^{2+} + 2SO_4^{\ 2-} + 16H^+ \quad \text{(EQ 10D)}$$

$$Fe_7S_{8(s)} + 17.25O_2 + 18.5H_2O = 7Fe(OH)_{3(s)} + 16H^+ + 8SO_4^{\ 2-} \text{ at pH} = 4 \quad \text{(EQ 11)}$$

$$FeSO_4 + H_2SO_4 + 0.5O_2 = Fe_2(SO_4)_3 + H_2O \quad \text{(EQ 12)}$$

$$FeS_{2(s)} + Fe_2(SO_4)_3 + 2H_2O + 3O_2 = 3FeSO_4 + 2H_2SO_4 \quad \text{(EQ 13)}$$

$$MS(s) + Fe_2(SO_4)_3 + 1.5O_2 + H_2O = MSO_4 + 2\,FeSO_4 + H_2SO_4$$
where MS(s) is a metal sulfide (EQ 14)

$$4Fe^{2+} + O_2 + 4H^+ = 4Fe^{3+} + 2H_2O \quad \text{(EQ 15)}$$

$$MnS_{(s)} + nFe^{3+} = Mn^+ + S + nFe^{2+} \quad \text{(EQ 16)}$$

The presence of manganese oxides at pH ≤ 5.5 also generates Fe^{3+} and can be as important in some environments as the reaction with bacteria (Evangelou 1995):

$$MnO_{2(s)} + 4H^+ + 2Fe^{2+} = Mn^{2+} + 2H_2O + 2\,Fe^{3+} \quad \text{(EQ 17)}$$

In some cases, acidic waters can also form by reactions between sulfide minerals and Fe^{3+} without oxygen, such as in groundwater (Rimstidt et al. 1994):

$$Fe^{2+} + 0.5O_2 + H^+ = Fe^{3+} + 0.5H_2O \quad \text{(EQ 18)}$$

$$FeS_{2(s)} + 14Fe^{3+} + 8H_2O = 15Fe^{2+}(aq) + 2SO_4^{2-}(aq) + 16H^+(aq) \quad \text{(EQ 19)}$$

$$PbS_{(s)} + 8Fe^{3+} + 4H_2O = 8H^+ + SO_4^{2-} + Pb^{2+} + 8Fe^{2+} \quad \text{(EQ 20)}$$

$$ZnS_{(s)} + 8Fe^{3+} + 4H_2O = 8H^+ + SO_4^{2-} + Zn^{2+} + 8Fe^{2+} \quad \text{(EQ 21)}$$

$$FeAsS_{(s)} + 13Fe^{3+} + 8H_2O = 13H^+ + SO_4^{2-} + 14Fe^{2+} + H_3AsO_4(aq) \quad \text{(EQ 22)}$$

$$CuFeS_{2(s)} + 16Fe^{3+} + 8H_2O = 16H^+ + 2SO_4^{2-} + Cu^{2+} + 17Fe^{2+} \quad \text{(EQ 23)}$$

$$Fe_2(SO_4)_3{\cdot}9H_2O_{(s)} = 2Fe(OH)_{3(s)} + 3SO_4^{2-} + 6H^+ + 3H_2O \quad \text{(EQ 24)}$$

$$FeAl_2(SO_4)_4{\cdot}22H_2O_{(s)} + 0.25O_2 =$$
$$Fe(OH)_{3(s)} + 2Al(OH)_{3(s)} + 4SO_4^{2-} + 8H^+ + 13.5H_2O \quad \text{(EQ 25)}$$

$$Fe^{3+} + 3H_2O = Fe(OH)_{3(s)} + 3H^+ \quad \text{(EQ 26)}$$

$$Al^{3+} + 3H_2O = Al(OH)_{3(s)} + 3H^+ \quad \text{(EQ 27)}$$

The most common acid-consuming reactions are with calcite and dolomite:

$$CaCO_{3(s)} + H^+(aq) = Ca^{2+} + HCO_3^- \text{ at pH} > 6.4 \quad \text{(EQ 28)}$$

$$CaCO_{3(s)} + 2H^+ = Ca^{2+} + H_2CO_3^0 \text{ at pH} < 6.4 \quad \text{(EQ 29)}$$

$$CaMg(CO_3)_{2(s)} + 2H^+ = Ca^{2+} + 2HCO_3^- + Mg^{2+} \text{ at pH} > 6.4 \quad \text{(EQ 30)}$$

$$CaMg(CO_3)_{2(s)} + 4H^+ = 2H_2CO_3^0 + Ca^{2+} + Mg^{2+} \text{ at pH} < 6.4 \quad \text{(EQ 31)}$$

Silicates may also consume acid at slower rates than carbonates by dissolution reactions:

$$KAlSi_3O_{8(s)} + H^+ + 7H_2O = K^+ + 3H_4SiO_4 + Al(OH)_{3(s)} \quad \text{(EQ 32)}$$

$$KAl_3Si_3O_{10}(OH)_{2(s)} + H^+ + 9H_2O = K^+ + 3H_4SiO_4 + 3Al(OH)_{3(s)} \quad \text{(EQ 33)}$$

$$CaAl_2Si_2O_{8(s)} + 2H^+ + H_2O = Ca^{2+} + Al_2Si_2O_5(OH)_{4(s)} \quad \text{(EQ 34)}$$

Cyanide leaching:

$$Au + 8NaCN + O_2 + 2H_2O = 4\,NaAu(CN)_2 + 4NaOH$$
(Elsner's equation, Adamson's first equation) (EQ 35)

$$2Au + 4NaCN + O_2 + 2H_2O = 2NaAu(CN)_2 + H_2O_2 + 2NaOH$$
(Adamson's second equation) (EQ 36)

Glossary and Abbreviations

This glossary includes terms related to sampling, monitoring, mitigation of mining influenced water, mining, and reclamation. All definitions given are common scientific usage except where a source is provided (in parentheses) following the entry. Full references are given at the end of the glossary under "Sources for Definitions." This glossary is a living document that will be added to or altered as additional volumes of the *Management Technologies for Metal Mining Influenced Water* series are completed. A current version of this glossary can be obtained at www.bucknam.com/glossary.doc.

ABA – See *acid–base accounting*.

abandoned mine – Excavations, structures, or equipment left from a former mining operation that, for practical purposes, have been deserted and upon which property there is no evident intent of further mining. An assumption of "abandoned" may be incorrect if an owner still exists, even if the owner has not performed any activity at the location for a long period, in which case the mine may simply be "inactive."

abandoned mine land (AML) – Lands that were mined and left unreclaimed where no individual or company has reclamation responsibility. These may consist of excavations, either caved in or sealed, that have been deserted and where further mining is not intended. In some jurisdictions, these are known as *legacy sites*.

abandoned workings – Excavations, either caved in or sealed, that are deserted and in which further mining is not intended; open workings that are not ventilated and inspected regularly. (*U.S. Bureau of Mines Federal Mine Safety Code*)

abatement – In regulatory or environmental cleanup language, action to reduce the degree or intensity of pollution or contaminant release, as used.

accuracy – (1) Measurement or statement comparing chemical analytical results to a true or accepted value; may be determined by analyzing certified standards as unknown samples and comparing with known certified values. (2) The closeness of approach of a measurement to the true value of the quantity measured. Because the true value cannot actually be measured, the most probable value from the available data, critically considered for sources of error, is used as the truth. (*American Society for Metals Glossary*)

acid–base accounting (ABA) – An ore, waste, and overburden analytical technique that determines the potential acidity from sulfur analysis, or more correctly, sulfide analyses, versus the neutralization potential of a rock or soil sample. It is used to predict the potential of that sample to be acid producing, neutral, inert, or acid neutralizing.

acid–base classes – Descriptive terms to identify materials according to their net calcium carbonate (NCC) percentage, acid neutralization potential (ANP), and acid generation potential (AGP), as follows:

Classification	Specifications
Highly acidic	NCC ≤ –10%
Acidic	–10% < NCC ≤ –2%
Slightly acidic	–2% < NCC ≤ –0.2%
Neutral	–0.2% < NCC < 0.2% and AGP ≤ –0.2% or ANP ≥ 0.2%
Inert	–0.2% < NCC < 0.2% and AGP > –0.2% or ANP < 0.2%
Slightly basic	0.2% ≥ NCC < 2.0%
Basic	2.0% ≥ NCC < 10%
Highly basic	NCC ≥ 10%

(ASTM E1915)

acid concentration present (ACP) – Test method to determine the acidity present in a sample by titrating a slurry of the sample with standards base.

acid drainage (AD) – Water with low pH and/or high acidity where such pH and/or acidity is attributable to weathering of sulfur or sulfide minerals; may derive from either natural weathering of sulfur or sulfides or acid weathering of mined deposits or waste rock.

acid generating – Refers to ore and mine wastes that contain sulfur or sulfides, which produce acid when oxidized. Acid can also be present as acid sulfates or generated by their weathering, produced originally from oxidation of sulfides, even when all sulfides have been oxidized in the wastes.

acid generation potential (AGP) – The potential of a sample to generate acid based on estimates of its sulfide sulfur concentration. (*ASTM E1915*)

acid mine drainage (AMD) – Acidic drainage generated by weathering of sulfur or sulfide minerals in mines, ores, or mine wastes within a mining area. AMD results from oxidation of some sulfide minerals exposed because of mining, which produces sulfuric acid. Such drainage typically contains high sulfate and dissolved iron from the weathering of iron sulfides. Typically, the sulfuric acid dissolves minerals in the rocks, which tend to release metals to the water. AMD implies that acid drainage is related to the mining process and does not include natural drainage.

acid mine water – (1) Mine water that contains free sulfuric acid, mainly due to the weathering of iron pyrites. A pit water, which corrodes iron pipes and pumps, usually contains a high proportion of solids per gallon, principally the sulfates of iron, chiefly ferrous and alumina. (*Nelson*) (2) Where sulfide minerals break down under chemical influence of oxygen and water, the mine drainage becomes acidic and can corrode iron work. If it reaches a river system, biological damage may also result. (*Pryor*)

acid neutralization potential (ANP) – The potential of a sample to neutralize acid based on estimates of its carbonate carbon potential concentration. (*ASTM E1915*)

acid potential (AP) – The ability of a rock or overburden to produce acid leachates; may be referred to as acid generation potential or AGP.

acid rock drainage (ARD) – A low pH, metal-laden, sulfate-rich drainage that occurs during land disturbance where sulfur or metal sulfides are disturbed. It forms under natural conditions from the oxidation of pyrite or other sulfide minerals and where the acidity exceeds the alkalinity. Nonmining exposures, such as along highway road cuts where sulfide materials are exposed, may produce similar drainage but technically are not mined areas. Also known as *acid mine drainage* when it originates from mining areas.

acid waste – Ore- or mineral-processing waste that contains sulfur, pyrite, or other sulfide minerals in excess of the neutralizing potential.

acidity – The titratable acid as measured in accordance with standard methods. It is normally reported as milligrams per liter as calcium carbonate ($CaCO_3$), but it may also be reported as milliequivalents per liter as hydronium (H^+).

ACMER – Australian Centre for Minerals Expansion and Research

ACP – See *acid concentration present.*

active mine – The area in which active mining takes place relative to extraction of metal ores, industrial minerals, and other nonrenewable natural resources of economic value.

active treatment systems – Systems that treat drainage with active addition of chemical reagents or the application of external energy to operate. One example is when basic chemicals such as hydrated lime [$Ca(OH)_2$], caustic soda (NaOH), or ammonia (anhydrous ammonia, or NH_3) are added to the water to neutralize acidity and to precipitate metals.

AD – See *acid drainage.*

adit – A horizontal or nearly horizontal passage driven from the surface for the working or dewatering of a mine. If an adit is driven through the hill or mountain to the surface on the opposite side, it is called a *tunnel.*

ADTI-CMS – Acid Drainage Technology Initiative–Coal Mining Sector

ADTI-MMS – Acid Drainage Technology Initiative–Metal Mining Sector

AGP – See *acid generation potential.*

agricultural grade limestone – Refers to ground limestone suitable for agricultural use, having at least 50% particles that will pass a 60-mesh (0.25-mm) screen.

agronomic – Dealing with growing of crops under cultivation.

agronomic grass – A grass species normally used in agriculture as forage or other crop.

agronomic rate – In reference to biosolids, a loading rate that is calculated, usually on a dry-weight basis, to provide sufficient mineralizable nitrogen to supply annual crop or vegetation needs. Biosolids applied at the agronomic rate typically means that the amount of nitrogen that passes below the root zone of the crop or vegetation is minimized.

ALD – See *anoxic limestone drain.*

alkaline trench – A trench packed with alkaline materials designed to neutralize, intercept, and convey surface acidic runoff into groundwater recharge.

alkalinity – The titratable alkalinity, using a standard acid titrant, as performed in accordance with standard methods. It is normally reported as milligrams per liter as calcium carbonate ($CaCO_3$), but it may also be reported as milliequivalents per liter as bicarbonate (HCO_3^-).

alluvium – Soil or sediments transported and deposited by flowing water.

aluminosilicate mineral – A mineral in rock or soil based on aluminum and silicon, such as a feldspar, mica, or clay mineral.

AMD – See *acid mine drainage.*

ameliorate – To improve or make better. Referring to soil, it improves soil with respect to its plant growth properties.

amendment – A material that is incorporated into the soil surface or subsurface horizons to improve soil quality and/or plant growth.

AML – See *abandoned mine land.*

analysis of variance (ANOVA) – A statistical procedure, used to assess whether significant differences are present between measured variables.

ANC – acid-neutralizing capacity

anion – An ion with a negative charge. An anion—such as chloride (Cl^-), nitrate (NO_3^-), bicarbonate (HCO_3^-), or sulfate (SO_4^{2-})—may result from the dissociation of a salt, acid, or alkali.

annual – A plant that completes its life cycle in less than 1 year.

ANOVA – See *analysis of variance.*

anoxic limestone drain (ALD) – A buried trench or cell of limestone into which anoxic water is introduced to raise pH and add alkalinity, without coating the limestone with ferric hydroxide [$Fe(OH)_3$].

ANP – See *acid neutralization potential.*

anthropogenic – Formed through or related to the activities of humans.

AOC – See *approximate original contour.*

AP – See *acid potential.*

appraisal – A process for determining the fair market value for property.

approximate original contour (AOC) – Surface configuration achieved by backfilling and grading of the mined area to the general configuration it had prior to mining so that highwalls are eliminated.

aquifer – A geologic formation, group of formations, or part of a formation that contains sufficient saturated permeable material to yield significant quantities of water to springs and wells.

ARD – See *acid rock drainage.*

area mining/stripping – Open cut or open pit surface mining that is carried on the level or along the prevailing topography in which relatively large tracts are disturbed. Usually mined with large shovels or draglines. See also *surface mining.*

aspect – The direction toward which a slope faces (e.g., south, northeast).

ASTM – ASTM International (formerly American Society for Testing and Materials)

B horizon – The illuvial soil horizon directly below the topsoil, which has been modified by soil forming, such as processes of accumulation.

backfill – Geologic materials returned to the pit or placed back into the mine, after desirable minerals have been removed, to bring a surface mine back to original contour, partially refill an open pit, or to improve stability of underground workings.

background – Natural concentrations of an element in natural materials that exclude human influence. A background measurement represents an idealized situation and is typically more difficult to measure than a *baseline*.

BAPP – See *biological acid production potential*.

bare-root stock – Tree seedlings from a nursery, usually at least 2 years old, to be planted with their roots bare of soil.

barren – Land that has little or no plant cover.

barrier pillars – Undisturbed ore left in place to support the roof between mine panels and to separate two adjacent underground mine properties. These panels can also protect adjacent underground mines from accidental flooding or explosions.

basal area – The sum of the cross-sectional area of all tree stems measured 1.4 m above the ground. Basal area is often used to estimate stocking levels after canopy closure.

base saturation – The percentage of cation exchange sites, usually expressed as centimole per kilogram (cmol/kg), that are occupied on a soil by cations (cmol/kg) other than hydrogen and aluminum.

baseline – A baseline measurement represents concentrations measured at some point in time and may or may not represent the true background. Baseline concentrations are typically expressed as a range, not as a single value.

bearing capacity – The amount of pressure or weight a soil or geologic material can withstand before failure—an important consideration for determining the suitability for building on that surface.

benched slope – A slope that has been constructed by making stair-like cuts into the side of a slope.

beneficial reuse – Using a substance normally considered a waste product in a new use or product.

beneficiation – The processing of ores for the purpose of regulating the size of a desired product; removing unwanted constituents; and improving the quality, purity, or assay grade of a desired product. Concentration or other preparation of ores can be for smelting by screening, drying, flotation, or gravity or magnetic separation. Improvement of the grade of ores can be by milling, screening, flotation, sintering, gravity concentration, or other chemical and mechanical processes.

benign neglect – An approach to disturbed site restoration that allows natural processes to play a major role.

benthos – The organisms living on or in the bottom of water bodies.

berm – A narrow bank of earth.

binders – Chemicals used to adhere fiber or paper mulch so the mulch remains stationary.

bioavailability – Degree of ability to be absorbed and ready to interact in organism metabolism.

biodiversity – The variety of living organisms at all levels of organization.

biological acid production potential (BAPP) – A kinetic testing procedure using a nutrient media and cultured bacteria to determine whether the biological acid production reaction can be sustained in the sample. (*ASTM E1915*)

biological magnification – The process in which certain substances, such as heavy metals, move up the food chain, work their way into rivers or lakes, and are eaten by aquatic organisms, such as fish, which in turn are eaten by large birds, animals, or humans. The substances become concentrated in tissues or internal organs as they move up the chain (synonym = biomagnification, bioconcentration).

biomass – Standing crop of living material usually expressed as the amount of live or dry weight per unit area; usually associated with soil microbes, animals, and plant residues.

biome – A major biotic unit consisting of plant and animal communities with similarities in form and environmental conditions.

biosolids – (1) An end product of municipal wastewater treatment plants or other organic waste materials. Commonly, sewage sludge is referred to as biosolids. (2) Sewage sludge that is placed on, or applied to the land to use the beneficial properties of the material as a soil amendment, conditioner, or fertilizer. Many regulatory definitions by the U.S. Environmental Protection Agency and various state regulatory agencies restrict the definition to beneficial uses and usually exclude such things as

- Sludge determined to be hazardous or special waste;
- Sludge with a specified concentration of polychlorinated biphenyls;
- Sludge generated during the treatment of either surface water or groundwater used for drinking water;
- Sludge generated at an industrial facility during the treatment of industrial wastewater, including industrial wastewater combined with domestic sewage; and
- Commercial septage, industrial septage, or domestic septage combined with commercial or industrial septage.

See also the Arizona Deptartment of Environmental Quality Web site, www.azdeq.gov/environ/water/permits/bio.html and EPA's rules (40 CFR 503) on the use or disposal of sewage sludge. (*Arizona Department of Environmental Quality*)

BLM – U.S. Bureau of Land Management

BNC – base-neutralizing capacity

bog – A peaty, acidic wetland, often hummocky, and always including species of *Sphagnum* (peat moss).

bog peat – Peat that is developed mainly from mosses of various *Sphagnum* species.

boreal – Northern, usually referring to the circumpolar zone of a coniferous forest.

borrow area – Place from which earthy materials are removed to serve as fill.

brackish water – Slightly salty water.

bridging voids – Voids formed in mine soils because overlapping rock fragments do not allow all spaces to be filled with fine-earth material during the backfilling process.

broadcast seeding – Scattering seed on top of the soil; commonly done with a hand seeder, aerial, or with a spreader mounted on a vehicle.

broadleaf plants – A group of plants that are dicots. They can be annual, biennial, or perennial and herbaceous or woody.

browse – Woody vegetation consumed by animals.

brush matting – Cut branches of small trees and shrubs laid on the soil surface during reclamation to prevent wind erosion and to protect establishing seedlings. In some circumstances, significant quantities of native species seed can be introduced in this way.

brushland plow – A heavily constructed, three-wheeled implement without a continuous, solid axle, designed to treat a 10-foot swath; each pair of discs is individually supported by a spring-loaded arm.

bulk density – A measure of the mass of soil or rock (or other solid phase material) per unit volume, g/cm^{-3}.

C horizon – A mineral horizon between the B horizon and bedrock that shows little effect of soil-forming processes.

calcareous soil – Soil containing sufficient free calcium carbonate ($CaCO_3$) and other carbonates to effervesce visibly or audibly when treated with cold one-tenth molar hydrochloric acid (0.1M HCl).

calcium carbonate equivalent (CCE) – A measure (ratio) of the relative neutralizing value of a liming material equivalence (CCE) in comparison to pure calcium carbonate ($CaCO_3$). For example, magnesium carbonate ($MgCO_3$) has a CCE of approximately 1.20 on a mass basis.

calibrated watershed study – A study in which the amounts, fluxes, and composition of ground and surface waters are monitored within a watershed or catchment area.

canopy cover – The percentage of the ground that would be covered if the crowns of trees and shrubs were projected vertically downward.

cap – See *covering material.*

carbonates – A family of rocks containing calcium (Ca) and/or magnesium (Mg) carbonate, such as limestone and dolomite, and which excludes siderite ($FeCO_3$).

carrying capacity – The ability of an ecosystem to support animal life. The term is often applied to human-managed systems to describe the limits at which use of a resource is considered sustainable.

catchment – See *drainage basin.*

cation – A positively charged ion; for example, sodium ion (Na^+) and hydronium ion (H^+).

cation exchange capacity (CEC) – The total cations that can be adsorbed onto or desorbed from soil particles, usually expressed as centimoles per kilogram (cmol/kg).

cation exchange site – A negatively charged site on a soil particle that is capable of attracting, holding, and exchanging cations.

CEC – See *cation exchange capacity.*

cement kiln dust – The alkaline dust resulting from the manufacture of cement, usually containing a high percentage of calcium oxide and hydroxide.

centibar – A measure of soil water potential.

chain pillar – A series of pillars left between panels that support the mine roof and allow access to the mine panels as well as air exchange in an underground mine.

chemisorbed – A substance that is chemically bound to the surface of another.

chlorophyll – The green pigment found in plants that is an agent of photosynthesis.

chlorotic – Refers to plant parts, especially leaves, having a yellowish or whitish appearance due to the nondevelopment or destruction of chlorophyll.

clay slime – The silt and clay fraction of mineral sand deposits released into the process water during mining. The slime interferes with separation of minerals from the sand mass, and its level must be reduced in the water for efficient mineral recovery.

clean tillage – Tillage in which no plant residue is left on the surface in contrast to no-till, in which seeds are planted into residue killed by herbicides.

cleanup – Actions taken to deal with a release or threat of release of a hazardous substance that could affect humans and/or the environment. The term is sometimes used interchangeably with remedial action, removal action, response action, or corrective action.

climax – The highest ecological development of a plant community capable of perpetuation under the prevailing climatic and edaphic conditions.

CMS – Coal Mining Sector (of ADTI)

colonization – The movement of new individuals or species into an area.

commercial forestry – Forest management that maximizes the economic value of tree stands for wood products.

community – Assemblages of plants and animals occurring in natural systems.

compaction – Increase in soil bulk density, reflected in increased penetrometer resistance caused by loading at the surface, generally by wheel traffic; the action of moving soil particles closer together by compressing the pore space.

composite sample – A sample made by the combination of several distinct subsamples. Composite samples are often prepared to represent a minable or treatable unit of material when it is not economically feasible or desirable to analyze a large quantity of individual samples; to represent a particular type or classification of material; or when subsample volumes are insufficient to allow analyses by desired analytical techniques. Metallurgical composites are normally prepared using subsample weights, which are proportional to their percentage in the ore so that the overall composite represents the ore grade of the material.

concentrating – The mechanical process, often involving flotation or gravity separation, by which the valuable part of an ore (i.e., concentrate) is separated from the gangue, or noneconomical rock minerals, to be further treated or disposed of as tailings.

concentrator – Part of the mineral-processing plant used to separate valuable minerals from the ore. In mineral sand mining, the concentrator is often referred to as a *wet mill* because it uses a water slurry for separation and is often floated on pontoons in a dredge pond.

coniferous forest – A forest dominated by evergreen trees species such as pines (*pinus*) and spruce (*picea*).

container stock – A nursery tree seedling, often less than a year old, that is grown in a container and planted while still in its growth medium.

contaminant – Any physical, chemical, biological, or radiological substance or matter that has an adverse effect on air, water, or soil.

contour mining – A mining operation that follows the ore along the contour of the mountainside or hillside.

cool-season plant – Plants that grow most rapidly under cooler air and soil temperatures, a C3 plant.

cover – The combined aerial parts of plants and mulch.

cover crop – A cereal grain (or other annual plants) used to prevent erosion and enhance organic matter prior to permanent seeding.

cover/frequency index – A modified importance value index obtained by placing percent cover and percent frequency, respectively, on a relative basis, then adding relative cover and relative frequency and dividing by 2.

cover soil – Soil topdressing placed over replaced subsoil or spoil material. Cover soil in the alpine and subalpine regions is composed of the salvaged organomineral surface soil material and the underlying B horizon material. See also *reclamation material.*

covering material – A material placed over waste material and often used as a plant growth medium; also called a *cap*.

crop trees – Commercially valuable tree species within a mix of species that are expected to dominate the canopy after closure.

cropland – Land used for production of adapted crops for harvest, alone or in a rotation with grasses and legumes. This type of land usually is the better farmland and is restored to the same productivity it had prior to mining.

crustaceans – Any member of the class *Crustacea* in the phylum *Arthropoda*. Examples of these include crabs, lobsters, and shrimp. Most species are aquatic but a few inhabit terrestrial environments.

cultipacking – A tillage tool in which a seedbed is prepared by packing below the depth of seed placement and leaving a loose or firm soil surface.

cut slope – A slope that has been constructed by the removal of soil or rock material.

data – Any information about a feature or condition; usually implies a numerical format for the data but also can be textual information. It also implies a standard method of collection and/or storage of such information.

data quality objective (DQO) – Qualitative and quantitative statement of the overall level of uncertainty that a decision-maker will accept in results or decisions based on environmental data. A DQO provides the statistical framework for planning and managing environmental data operations consistent with the user's needs.

decant zone – A location farthest downslope from the discharge point in a tailings pond, where supernatant water is removed from above-settled tailings materials.

deciduous forest – A forest dominated by trees that lose their leaves each autumn.

delta – An accumulation of sediment at a river mouth.

denitrification – When soils become saturated, oxygen is excluded, or water becomes deoxygenated, and anaerobic decomposition takes place. Anaerobic and facultative microorganisms have the ability to obtain their oxygen from soil or soluble nitrates and nitrites, resulting in the net loss of nitrogen gas (N_2) from the soil or water system into the atmosphere.

density – The number of individuals per unit area.

denuded land – Land from which vegetation and soil has been removed (in earth science, refers to soil as a result of erosion).

detention time – The time in hours that runoff is in a sediment pond, to allow sedimentation prior to its removal through a spillway.

detoxify – Referring to soil, to render a phytotoxic soil less toxic.

detritus – Any dead or decaying organic debris derived from plants.

development rock – Rock materials removed in the development of a mine or ore body; may or may not contain recoverable minerals.

dieback – A phenomenon in which trees progressively lose more and more of their living branches. In the western United States, it also refers to a decrease in vegetation cover after irrigation has been terminated.

direct seeding – Sowing seeds by broadcasting onto the surface of prepared sites.

discharge point – Location at which mineral-processing waste is pumped into a basin or impoundment.

disc plowing – Plowing with discs for controlling small, shallow-rooted plants and for preparing a seedbed.

discing – The use of a piece of agricultural machinery, called a disc, in which sharp-edged, angled, metal discs are drawn across the land as a way of breaking up the soils, incorporating amendments, and killing unwanted vegetation.

dissolved oxygen – A measure of water quality indicating oxygen dissolved in water. This is one of the most important indicators of a water body's condition, because dissolved oxygen is necessary for the life of fish and most other aquatic organisms.

dissolved solids – The weight of matter, including both organic and inorganic matter, in solution in a stated volume of water. The amount of dissolved solids is usually determined by filtering water through a glass or 0.45-pore-diameter micrometer filter, weighing the filtrate residue remaining after the evaporation of the water, and drying the salts to constant weight at 180°C.

diversity – The magnitude and evenness of the species composition of plant communities.

dolomitic limestone – Limestone (calcium carbonate) containing a significant percentage of dolomite (calcium-magnesium carbonate).

dozer – A bulldozer or tractor equipped with a blade or steel plate on the front end so that it can be used to push earthen spoils primarily in the forward direction. It may be on tracks or rubber tires.

dozer basin – A large basin gouged with a bulldozer blade, made for the purpose of trapping precipitation and increasing infiltration.

DQO – See *data quality objective.*

dragline – An excavating machine that uses a bucket operated and suspended by cables and a boom. The bucket is dragged back toward the machine along the surface and usually operates from the highwall or above the area in which soil or spoil is being removed.

drainage – Any water draining from a natural or human-made feature, including natural surface water runoff, mine drainage, and groundwater that has come to the surface.

drainage basin – The surface between topographic divides that receives precipitation; also known as a *catchment* or *watershed.* This water is conveyed downslope as surface runoff or groundwater.

dredge mining – Frequently used in mineral sand mining to recover the ore from the mine face as a water slurry by suction. The dredge houses a powerful pump and is floated on pontoons in the dredge pond. In a suction cutter dredge, ore recovery is assisted by a revolving open basket (cutter) mounted over the suction inlet. In hard ground, the cutter can be replaced by a rotating underwater bucketwheel. This term can also be applied to placer mining.

dredged materials – Soil materials that have been removed from (shipping) channels of waterways by dredging operations and deposited elsewhere.

drill – An implement used in agriculture and reclamation to place grass, forb, shrub, and tree seeds into the soil. Also a machine with a rotating and/or hammer-driven bit used to drill holes in rock units. In the case of hard-rock mining, these holes may be partially filled with explosives for blasting and loosening the rock to be removed in mining,

ecological processes – Climate, topography, and soil factors interacting with the flora and fauna.

ecological sink – The development of isolated patches of habitat that attracts animals in greater numbers than the habitat can support.

ecology – The study of the interrelationship of organisms with their environment.

ecosystem – A community of organisms considered together with the nonliving factors of its environment.

ecotone – An area where two major communities meet and blend together.

ecotype – A population or group of populations of a plant species, distinguished by genetically based morphological or physiological characteristics, which has developed in response to specific environmental factors.

effluent – A material, usually a liquid waste, that is emitted by a source, which is often industrial, such as a metallurgical or water treatment process. Gaseous effluents are usually called *emissions.*

Eh – A measure of the reduction–oxidation status of soil–water systems and metallurgical processes. An oxidizing environment has a positive Eh, and a reducing environment has a negative Eh, relative to a standard reference electrode.

electrical conductivity – Indicates the concentration of ionized constituents in a water sample or soil matrix.

emissions – Gaseous materials emitted by a source.

enclosure – The sense of being enclosed, surrounded, or contained by elements of a landscape design, as in a courtyard, a clearing in a wooded area, or a small valley. It implies bringing a larger landscape to a more personal or human scale to facilitate various activities. It can include a fence to retain or prevent grazing of livestock or wildlife in a specific area.

environmental impact – A process required under the U.S. National Environmental Protection Act for project assessment involving federal or state money, in which potential physical and social impacts and mitigation measures are discussed and analyzed. A provision for notifying citizens and considering their comments is integral to the process.

EPA – U.S. Environmental Protection Agency

ephemeral stream – A stream that flows only during and after a storm or snowmelt.

erosion – The entrainment and transportation of soil through the action of wind, water, or ice.

estuary – The area where a river meets the sea, usually in a body of water transitional in size between the river and the sea, with a resulting mixture of saline and freshwater. These areas possess characteristic flora and fauna.

exotic – Not native to an area or ecosystem.

extraction – The process of mining and removal of ore from a mine. This term is often used in relation to all processes of obtaining metals from ores, which involve breaking down ore both mechanically (crushing) and chemically (decomposition), and separating the metal from the associated gangue.

extraction ratio – The ratio of the amount of ore removed to the amount of ore remaining in a mine or disposed of as waste.

fauna – Animal component of natural systems.

fertilizer grade – Guaranteed minimum (in percent) of the major plant nutrients contained in a fertilizer. For example, a fertilizer with a grade of 20-10-5 contains 20% nitrogen (N), 10% available phosphate (P_2O_5) or 4.37% phosphorus (P), and 5% available potassium oxide (K_2O) or 4.15% potassium (K). Micronutrients may also be included.

fill slope – A slope that has been constructed by the addition of unconsolidated earth materials.

filter cloth – Plastic or fiber cloth that water can pass through but most sediment cannot. It is used in constructing sediment fences and riprap filters.

fine tailings – Fine-grained clastic materials (silts and clays) that consolidate very slowly during settling of tailings.

finish grading – A slope-grading technique in which the surface is left smooth and uniform and free of exposed rock, soil clods, or most undulations.

floc – An agglomeration of very fine particles within water caused by their chemical or charge affinity for each other. Flocs are meant to settle out of the water faster than the individual particles because they are heavier in agglomerated form but sometimes may float on the surface because of entrained gasses. The metal hydroxide precipitates and sludges collected in settling ponds after neutralization, coagulation, and flocculation treatment are sometimes referred to as floc.

flora – The plant component of natural systems.

flotation – The method of mineral separation in which a froth, created in water by a variety of reagents, floats some finely ground minerals while other minerals sink.

flume pipe – A length of pipe used to channel water. In pipeline construction, the stream is blocked off on the upstream side of a right-of-way and stream water is channeled into the pipe over or under a right-of-way and discharged back into the stream on the downstream side of a right-of-way. This keeps water out of the pipe ditch while it is open.

focal point – An element in a design to which attention is deliberately drawn by the use of other elements of a design. Examples of focal points include statues, building entrances, a special tree, or a fountain.

fool's gold – See *pyrite.*

footprint – The planimetric area covered by a mine operation and associated roads, ponds, and other structures.

forest land – Land previously in a forest and returned to this land use following mining.

forest peat – Peat that is composed of tree and shrub remnants mixed with mosses.

forest productivity – Forest biomass production per unit area per unit time (e.g., $m^3/ha^{-1}/yr^{-1}$).

fracture flow – The flow of fluid, usually water or including water, through fractures in geological media (rock units or indurated sediments). Fractures may represent the dominant flow paths through media that otherwise are relatively impermeable, such as for igneous rocks and high clay-content units.

framboidal pyrite – Spherically shaped agglomerations of minute (approximately 0.25 µm) crystals of pyrite (FeS_2). It is the most reactive of all pyrite morphologies.

frequency – In an ecological context, a quantitative expression of the presence or absence of individuals of a species in a population.

gangue – The valueless minerals in an ore; that part of an ore that is not economically desirable but cannot be avoided when mining the deposit. It is separated from the ore during beneficiation.

geoavailability – That portion of a chemical element's or compound's total content in an earth material that can be liberated to the surficial or near-surface environment (or biosphere) through mechanical, chemical, or biological processes. The geoavailability of a chemical element or compound is related to the susceptibility and availability of its resident mineral phase(s) to alteration and weathering reactions. (*Smith and Huyck*)

geographic information system (GIS) – A computer program or system that allows storage, retrieval, and analysis of spatially related information in both graphical and database formats.

geostatistics – The mathematical assessment of variability in a biological, chemical, or physical parameter across a distance or area.

geotextile – A natural or synthetic fabric blanket that is permeable to fluids and gases.

GIS – See *geographic information system.*

glacial till – A heterogeneous mixture of clay, silt, sand, gravel, and boulders left behind by a glacier.

gouging – A soil preparation technique that produces deep, roughly oval depressions. An adapted chisel plow is used to form the depressions.

grab sample – A single sample collected at a particular time and place that represents the composition of the water, air, ore, waste, or soil only at that time, place, and of the quantity of material collected. It may be considered an opportunistic or typical sample and may not be representative of the overall material being sampled.

gradient – The inclination of profile grade line from the horizontal, expressed as a percentage (synonym = rate of grade).

grass crown – The position on the grass plant stem where two or more nodes stay close together and the secondary root system and tillers develop.

grass tiller – The asexual development of a new stem of a grass plant from auxiliary meristematic tissue of the parent plant.

grassland – A designated land use for reclamation on which the vegetation is dominated by grasses.

grid – A regular layout, usually square or rectangular, of locations for sampling or plots for studies of local conditions (e.g., revegetation plots and application of soil amendments).

grid sample – A single sample collected at a grid point at a particular time that represents the composition of the water, air, ore, waste, or soil only at that time, place, and of the quantity of material collected. Because of repeatability of surface conditions, it is generally preferred for systematic sampling.

grooved – A soil preparation technique, which creates 8-to-15-cm-deep parallel rips in the soil surface parallel to the slope face. The grooves are spaced 38 to 60 cm apart.

groundwater – Water in the zone below the surface of the earth where voids are filled with water and the pressure is 9.9 MPa (1 atm). This is in contrast to surface water.

growth medium – A material in which plants can be grown, usually soil or soil-like.

habitat – The place where a population (e.g., human, animal, plant, microorganism) lives, its surroundings, and its contents, both living and nonliving.

habitat patch – An area of consistent landscape and species composition, which meets some minimum size and shape requirements to support or attract a species of interest.

hardwood – A forester's term for a broad-leaved, deciduous tree with relatively dense woody tissue.

hardwood draw – See *woody draw.*

harrowing – Use of harrows to smooth rough plowed soil, control weed seedlings, and further seedbed preparation.

hay land – Land used to produce forage that is cut, dried, and stored for later use by animals.

heavy mineral sands – Valuable minerals such as rutile, ilmenite, leucoxene, zircon, and monazite occurring as a sand-sized fraction, with a high specific gravity relative to that of the host sand, usually silica.

hematite – A type of iron ore and mineral with the composition formula of Fe_2O_3.

herbaceous – Nonwoody plants or plant materials.

herbicide – A chemical preparation, often selective in its action, used to kill or inhibit the growth of plants.

heterotrophic bacteria – Bacteria that depend on palatable organic matter as their energy source and, therefore, are involved in the decomposition of leaf litter, and so forth.

high dune – A massive sand mass, usually more than 30 m and often in excess of 100 m in height, associated with present-day or relic coastal systems, and usually vegetated with heath, scrub, or forest ecosystems. It may host heavy mineral sands ore bodies formed by wind action.

higher or better use – A real-estate term referring to a land use having a greater financial return.

Hodder gouger – A piece of specialized reclamation equipment, originally designed by Richard Hodder of Montana State University, to modify microtopography of the soil surface. Normally towed behind a tractor, a gouger consists of three to five semicircular heavy steel blades attached to solid arms, which are attached to a heavy-duty frame; blades are arranged in one or two rows. The frame is supported with side wheels that are periodically raised and lowered to use the blades to scoop out depressions in the soil surface. The unit is operated hydraulically and has positive depth control and automatic up-and-down cycling. A seed box may be attached.

hoedad – A tree-planting tool shaped like a heavy-duty hoe.

hollow – See *valley-head hillslope*.

hot spot – An area of ore, mine soil, spoil, tailings, or waste with a pH <4.0, commonly encountered in mining areas where isolated areas of pyritic materials are exposed to weathering at the surface.

hydraulic sluicing – A method of mineral sand mining in which the ore is washed from the mine face by a strong water jet (monitor) and the resultant slurry is collected in a sump and pumped to the concentrator.

hydric soil – Soil that is saturated for a prescribed period during the growing seasons and is used as one of the indicators of a wetland.

hydrocarbons – Organic chemical compounds of hydrogen and carbon atoms that form the basis of all petroleum products.

hydrologic system – Composed of surface, underground, and atmospheric water, and the cycles between them in a given area.

hydrolysis – The process of splitting the water molecule into separate components of hydronium ions (H^+)and hydroxide ions (OH^-) that often react with other constituents present.

hydrolyze – A chemical reaction involving the water molecule and an element, usually a cation.

hydromulch – A mixture of mulch and water that may contain seed and fertilizer and is then sprayed onto a site.

hydroseeding – Dissemination of seed hydraulically in a water medium through a high-pressure nozzle. Mulch, lime, and fertilizer can be incorporated into the sprayed mixture. Hydroseeder is the trademark name for a specific type of hydraulic seeder.

ICARD – International Conference on Acid Rock Drainage

impoundment – A closed basin that is dammed or excavated and is used for the storage, holding, settling, treatment, or discharge of water, sediment, and/or liquid wastes.

imprinting – Increasing and stabilizing land productivity through improved control over rainwater infiltration by using a drum. The drum is usually filled with a liquid (e.g., water), and its surface has raised features with complex geometric patterns. This imprints the complex features in the soil when rolled across the surface. See also *land imprinter.*

inactive mine – A mine site in which no active mining is currently taking place relating to extraction of metal ores, industrial minerals, and other minerals of economic value. This is different from an abandoned mine, where there is no intent for further mining.

INAP – International Network for Acid Prevention

induction period – The time required for fine tailings to settle before a distinct interface is seen separating the fine materials from the clarified released water.

infiltration – The downward entry of water into a soil or other geologic materials.

infrastructure – Elements that support development, including transportation, utility, and communication systems.

inoculum – A source or medium for introduction of symbiotic microorganisms such as *Rhizobium*, often coated onto legume seeds.

in-stream flow devices – Synthetic structures that alter stream flow patterns.

intermediate water body – A water body that is 3 to 30 m (10 to 100 ft) in width.

interseeding – Seeding into an established vegetation cover. Often seeds are planted into the center of narrow seedbed strips of variable spacing and prepared by mechanical or chemical methods.

intertidal zone – The area between low and high tides.

interveinal chlorosis – Yellowish or paler green streaks between the veins of leaves, where the area close to the vein remains green.

introduced species – Plant and animal species that are not native to a certain area but have been introduced by humans, either deliberately or accidentally.

ion – An atom or group of atoms having an electrical charge, resulting from the gain or loss of one or more electrons from the elemental state.

island reclamation – Involves the placement of subsoil and topsoil in a pattern of larger, irregular-shaped areas on regraded spoil.

jack dams – A low-level water control structure constructed within the stream channel.

jarosite – A pale yellow to gray-green potassium iron sulfate mineral [$KFe_3(OH)_6(SO_4)_2$] that forms under active acid sulfate conditions; can be a pathfinder mineral for areas of oxidation of iron sulfides and associated acid generation. Jarosites may or may not contain acidic sulfate salts.

K-mica – An aluminosilicate mineral in which two silica tetrahedral sheets alternate with one octahedral sheet, with entrapped potassium atoms fitting between the sheets.

kriging – A geostatistical analysis procedure used to estimate an ore grade or biological, chemical, or physical value at locations where no samples were collected.

land imprinter – A specialized piece of reclamation equipment that creates a series of geometric impression (depressions) patterns on the soil surface. It is towed, normally has two drums with different imprint patterns on them on a common axle, and rolls over the soil surface, creating impressions in the soil to control erosion, improve the seedbed, and generally modify surface soil microtopography. A broadcast seeder can be mounted on the back of the frame.

land use – The primary use of a specific land area.

landscape inventory – The mapping, cataloging, or listing of landscape features, conditions, and influences on the landscape.

leaching – Removal by dissolution, adsorption, absorption, or substances, or chemical reaction from a matrix by passing liquids through the material.

leaf internode – The stem of a plant between leaf attachments or nodes.

leaf litter – The layer of dead leaves that forms under a plant cover, especially forests.

leaf node – The region on a stem where the leaf of a plant is attached.

legacy site – See *abandoned mine land.*

legume – A member of the pea family (*Leguminosae* or *Fabaceae*).

lethal – Deadly or fatal. Lethal dose is the amount of a toxic substance required to cause death of an organism under study in a given period of time.

lichen – A dual organism formed by a fungus and a green alga or cyanobacterium (blue-green alga) living together symbiotically.

lift – An engineering term for a layer of ore, overburden, or waste materials stacked in leach pads or waste dumps.

lime requirement – The amount of alkaline material needed to attain a desired soil pH, usually 6.5–7.0. This value is usually expressed in terms of calcium carbonate ($CaCO_3$) equivalency.

limestone – A sedimentary rock consisting largely of calcite ($CaCO_3$).

linkage – A connection or interface between places or land uses.

lithology – The character of a rock described in terms of its structure, color, mineral composition, grain size, and arrangement of its component parts—visible features that in the aggregate impart individuality to the rock. The term is often used to classify rock materials for characterization purposes along with the degree of alteration and acid–base characteristics.

litter – A layer of dead plant material on the soil surface.

littoral – Shallow water zone in lakes and ponds.

lysimeter (zero tension) – An enclosed column containing a solid (such as soil) that is used to measure leaching and/or evapotranspiration.

macronutrient – An element or compound required in relatively large amounts by an organism (e.g., N, P, K, Ca, Mg, and S in the case of green plants).

magnetite – A magnetic iron mineral that has the formula Fe_3O_4 and can form iron ore.

major water body – A water body that is more than 30 m (100 ft) in width.

management zone – The unique (physical and chemical) portion of substrate within coal slurry impoundments.

manure – Solid wastes from livestock, often mixed with bedding materials such as straw or hay.

matrix – A mixture of phosphate pebbles, sand-sized phosphate and quartz particles, silt, and clay. The proportions within the matrix mixture will differ among mining regions.

maximum contaminant level (MCL) – The maximum concentration of specific contaminants that is allowed in drinking water or in a permitted discharge under the U.S. Safe Drinking Water Act and other federal, state, and local regulations.

maximum potential acidity (MPA) – A value used in the acid–base account to evaluate the total amount of acidity from a rock sample that may result upon complete oxidation of the total amount of sulfur. The percent S is multiplied by 31.25 to yield acid in milligrams $(1{,}000\ mg)^{-1}$ (tons per 1,000 tons) of the material. This method is not normally applicable to metal mine ores and wastes nor to metallurgical processing products, but is applicable only when all of the sulfur is present as potentially acid-generating sulfides.

MBF – See *thousand board feet.*

MCL – See *maximum contaminant level.*

meander – Curve or bend in a stream channel.

MEND – Mine Environment Neutral Drainage

metal-complexing capacity – The ability of a material to combine with metal ions, sometimes rendering the metals unavailable to plants and becoming nontoxic.

metal tolerant – Plants that are able to grow in growth media containing levels of toxic metals that would preclude the growth of other or "normal" plants.

metallurgy – The science and technology of extracting and refining metals and the creation of materials or products from metals.

metallurgical processing – The methods employed to clean, process, and prepare metallic ores for the final marketable product.

microclimate – Localized temperature, moisture, wind, and other climate conditions, caused by differences in hydrologic climate and vegetation differences due to topography or surface configuration. For example, windbreaks create microclimates.

microcosm – A miniature world ecosystem.

microhabitat – A small-scale habitat influenced by local conditions.

micronutrient – An element or compound required in relatively small amounts by an organism (e.g., B, Cu, Zn, Mn, and Mo in the case of green plants).

milling – The crushing and grinding of ore. The term may include the removal of valueless or harmful constituents from the ore and preparation for additional processing or sale to market.

MiMi – Mitigation of the Environmental Impact from Mining Waste

mine – An opening or excavation in the ground for the purpose of extracting minerals.

mine mix – A 1:1 by volume mixture of composted sludge cake mixed with wood chips as a bulking agent and anaerobically digested dewatered sludge cake. Mine mix is produced by the city of Philadelphia, for example, for the purpose of mined land reclamation.

mine soils – Soils constructed and developed after the surface removal of minerals from the earth.

mineral deposit – An occurrence of any valuable commodity or mineral that is of a sufficient size and grade (concentration) that might under favorable conditions have potential for economic development.

mineral nutrient – An inorganic nutrient used by green plants.

mineral soil – A soil consisting predominantly of, and having its properties determined predominantly by, inorganic materials. Such soil can, however, contain up to 20% organic matter.

mineralogy – The study of minerals and their formation, occurrence, use, properties, composition, and classification; also refers to the specific mineral or assemblage of minerals at a location or in a rock unit.

minimal amelioration – Improving the condition of a growth substrate to the point at which plants can grow but not beyond this point.

mining – The process of extracting useful minerals from the earth's crust.

mining district – A section of country usually designated by name, having described or understood boundaries within which mineral deposits are found and mined under rules and regulations prescribed by the miners therein or by a government body. There is no limit to its territorial extent, and its boundaries may be changed if vested mineral or property rights are not interfered with. A mining district can be either an informal name for a mineral area or a legally defined area encompassing all or part of a collection of mineral deposits and/or mines.

mining influenced water (MIW) – Water sources that have been affected by mining and/or mineral-processing activities, including alkaline and neutral drainage, acid drainage, acid rock drainage, acid mine drainage, process solutions containing metal lixiviants (e.g., cyanide, acids, bases), and the degradation products of process solutions such as those with elevated ammonia (NH_3), cyanate (CNO^-), nitrate (NO_3^-), and thiocyanate (SCN).

minor water body – A water body that is 3 m (10 ft) or less in width.

mitigation – Correction of damage at the surface caused by mine subsidence, wetland impacts, acid drainage, and so forth.

MIW – See *mining influenced water.*

MMS – Metal Mining Sector (of ADTI)

modulus of rupture – A measure of the ability of a substance to break or resist breakage.

monitoring – The periodic or continuous surveillance or testing to determine the level of compliance with process or statutory requirements in various media or in humans, plants, and animals.

monospecific stand – A community of plants consisting almost entirely of a single species.

morphological – Dealing with the form and structure of hillside, stream channels, soil pedon, or organism, with special emphasis on external features.

mosses – A group of nonvascular green plants, sometimes sharing a similar ecological role in plant communities to the fundamentally different lichens.

MPA – See *maximum potential acidity.*

mulch – A layer of nonliving material applied or occurring on the surface that is placed on top of a growth medium to control erosion and weed growth and to conserve moisture.

mulch crop – A crop grown specifically for its organic materials, which are left on the soil surface to protect the soil and tree roots from the effect of raindrops, wind erosion, soil crusting, freezing, and evaporation.

multiple land use – Using a specific land area for more than one use at any specific time.

muskeg – A water-saturated form of peat; a term commonly used at high latitudes.

mycorrhiza (fungus root) – The symbiotic relationship between higher plants such as trees and fungi, in which trees benefit from enhanced water and nutrient uptake, while fungi benefit from photosynthesized products supplied by the tree.

native species – Plant or animal species that are indigenous to an area.

natural succession – Progression of a series of biotic communities over time.

NCC – See *net calcium carbonate.*

NCV – net carbonate value. See also *net calcium carbonate.*

neotropical migrants – Bird species that breed in the Northern Hemisphere and winter in the tropics.

nesting structures – Human made structures such as wood duck and bluebird nest boxes to supplement or replace natural nesting structures.

net calcium carbonate (NCC) – The result of the acid base account using sulfide sulfur and carbonate carbon estimates of acid neutralization potential and acid generation potential expressed in units of percent calcium carbonate. (*ASTM E1915*) NCC may also be expressed as net carbonate value (NCV) in units of percent carbon dioxide.

net calcium carbonate (NCC) classes – Descriptive terms to use for ranges in NCC as follows:

Classification	Specifications
Highly acidic	NCC ≤ –10%
Acidic	–10% < NCC ≤ –2%
Slightly acidic	–2% < NCC ≤ –0.2%
Neutral	–0.2% < NCC < 0.2% and AGP ≤ –0.2% or ANP ≥ 0.2%
Inert	–0.2% < NCC < 0.2% and AGP > –0.2% or ANP < 0.2%
Slightly basic	0.2% ≥ NCC < 2.0%
Basic	2.0% ≥ NCC < 10%
Highly basic	NCC ≥ 10%
(*ASTM E1915*)	

net carbonate value (NCV) – See *net calcium carbonate.*

net neutralization potential (NNP) – The result of the acid–base account that quantifies the amount of excess neutralization potential (positive result) or acid generation potential (negative result) in a sample in units of parts per thousand as calcium carbonate (tons $CaCO_3$/1,000 tons).

neutralization potential – The amount of alkaline or basic material in rock or soil materials that is estimated by acid reaction followed by titration to determine the capability of neutralizing acid from exchangeable acidity or pyrite oxidation. May be referred to as *acid neutralization potential* or ANP.

niche – The functional role of a species in the community, including activities and relations.

nitrogen fixation – The ability of certain microorganisms and plant–microorganism associations to convert atmospheric nitrogen into organic nitrogen compounds.

nitrogen mineralization – The conversion of organic soil nitrogen compounds (unavailable to plants) into inorganic (plant available) forms of nitrogen via biologically mediated processes.

NNP – See *net neutralization potential.*

nodulation – The formation of nitrogen-fixing swellings on the roots of plants that are able to fix nitrogen in association with symbiotic bacteria (e.g., legumes and alders).

nonpoint source – Diffuse discharge sources (e.g., without a single point of origin or not introduced into a receiving stream from a specific outlet). The discharges are generally carried off the land by storm water or through groundwater flow and seepage. Common nonpoint sources are agricultural storm water, return flows from irrigated agriculture, forestry, urban, mining, construction, dams, channels, land disposal of wastes, saltwater intrusion, and city streets.

nose slope – See *spur-end hillslope.*

N-P-K fertilizer – A fertilizer containing nitrogen, phosphorus, and potassium, in which the relative proportions of the three main constituents are expressed in a formula such as "6-24-24."

nucleation – The ability of certain plant species to form a nucleus for the spread of other species.

nurse crops – Plant species that are used to establish or to temporarily stabilize soil, protect it from wind, hold snow, and provide shade or nitrogen. As the species become established, they readily yield to the perennial forage species.

ochric epipedon – A surface horizon of mineral soil that is too light in color, too high in chroma, too low in organic carbon, or too thin to be any other epipedon described in the U.S. system of soil taxonomy; or that is both hard and massive when dry.

offset discing – Uses of two or more gangs of discs, each gang on a separate axle and frame, and arranged so that soil is moved in two directions.

OLC – See *open limestone channel.*

oolitic – A rock consisting of small round grains, usually of iron oxide or calcium carbonate ($CaCO_3$), cemented together.

open-bed roasting – The heating of a sulfide ore in the open, to drive off some of the sulfur in the form of sulfur dioxide (SO_2) gas and to oxidize the iron in preparation for smelting.

open limestone channel (OLC) – A surface channel or ditch filled with limestone that is used to convey and treat surface water and acid drainage.

opportunistic sample – Usually a random sample, taken at a location and time of convenience to the sampler and representative only of conditions at that time, place, and for the quantity of material collected. Results from such samples may be neither representative nor reproducible.

ore – The naturally occurring material from which a mineral or minerals of economic value can be extracted profitably or to satisfy social or political objectives. The term is generally, but not always, used to refer to metalliferous material and is often modified by the names of the valuable metal constituents.

ore deposit – A mineral deposit that has been tested and found to be of sufficient size, grade, and accessibility to be extracted for a profit at a specific time, based on economic assumptions.

organic matter – The accumulation of disintegrated and decomposed biological residues and other organic compounds synthesized by microorganisms or used in mining and metallurgical processing; found in soils, ores, concentrates, waste rocks, tailings, and metallurgical processing wastes.

organic slime – Fine organic material, usually colloidal, released into the process water during mining of heavy mineral sand ore deposits, usually those that are cemented by humate (spodic horizons). The slime interferes with the separation of the minerals from the sand mass, and its level must be reduced in the water for efficient mineral recovery.

organic wastes – Organic waste products from various industries; may include biosolids. Examples include sewage sludge, poultry manure, horse bedding, hog manure, and wood products such as sawdust, paper wastes, bark, and spent activated carbon.

orphan mined land – Unreclaimed land that was mined before state or federal laws required reclamation, with no private owner held responsible for reclamation of the mined land; same as abandoned mine land.

orthents – Entisols that have either textures of very fine sand or finer particles in the fine-earth fraction, or textures of loamy fine sand or coarser; a coarse fragment content of 35% or more; and an organic carbon content that decreases regularly with depth. These soils are not saturated with water for periods long enough to limit their use for most crops. Orthents is a suborder in the U.S. system of soil taxonomy.

OSM – U.S. Department of the Interior, Office of Surface Mining Reclamation and Enforcement; also known as OSMRE.

outslope – Steeply sloping areas located around the perimeter of a surface mine, overburden or waste dump, spoil pile, mill tailings embankment or dam.

overburden – Material of any nature, consolidated or unconsolidated, that overlies a deposit of useful and minable materials or ores, especially those deposits that are mined from the surface by open cuts or pits.

oxidation – A chemical process involving a reaction(s) that produces an increase in the oxidation state of elements such as iron and sulfur.

oxidize – The chemical reaction involving the removal of electrons from an element or compound [e.g., Fe(II) $\rightarrow$ Fe(III)].

oxidized zone – That part of the soil–geologic column from which sulfide minerals have been completely oxidized away, compared with the reduced zone; often equivalent to the *zone of weathering*.

PAG – See *peroxide acid generation*.

paleoenvironment – The ancient geologic setting (climate, geography, etc.) under which strata were deposited.

panel – A roughly rectangular development and extraction area in an underground mine where mining takes place. Panels are usually bounded by chain or barrier pillars.

papermill sludge – A by-product of the paper industry consisting mainly of ligno-cellulosic materials discarded during the papermaking process.

particle density – The mass per unit volume of soil or mineral particles.

particulate emissions – Solid particles and liquid droplets emitted into the atmosphere.

passive recreation – Generally informal, unorganized recreational activities of individuals or small groups, which have a low impact on the land, requiring very little modification of the land and few or no structures (e.g., walking, stargazing, and picnicking).

passive treatment systems – Systems that treat acid mine drainage without continual additions of chemicals, including aerobic and anaerobic wetlands, anoxic limestone drains, successive alkalinity-producing systems, and open limestone channels.

pastureland – Grazing lands, planted primarily with introduced or native forage species that receive periodic renovation and/or cultural treatments such as tillage, fertilization, mowing, weed control, and irrigation.

pathway – The physical course a chemical or pollutant takes from its source to an exposed organism.

PCB – polychlorinated biphenyl

peatland – Terrain such as bogs and fens, in which the soil is predominantly peat, a deposit of partly decomposed plant material formed under waterlogged conditions.

penetrometer – A device used to measure the resistance of soil to penetration, and a measure of soil strength or compaction.

percent cover – The percentage of the ground that would be covered if all parts of a particular species or the vegetation in a sample plot were projected vertically downward.

percent frequency (species) – The percentage of random sample plots in which a particular species appears.

perennial – A plant that lives from year to year, for 3 years or more. May also refer to a stream that flows all year.

performance bond – Surety, collateral, self-bond, or combination by which a permittee assures faithful performance and adherence to mining regulations.

performance standards – Standards that describe measurable conditions that a project must meet (e.g., the pH of water leaving the site must be between X and Y) without dictating the method of attaining the condition.

permanence – The expected length of longevity without renovation or reseeding.

peroxide acid generation (PAG) – A test method used to determine the presence of an excess of reactive sulfides in a test sample by mixing the sample overnight with a hydrogen peroxide solution. (*ASTM E1915*) It may be used to eliminate false positive biological acid production potential results for inert materials.

persistent species – Perennial species that are adapted to a region and persist with minimal input after they are established.

pH – A measure of the acidity (less than 7) or alkalinity (greater than 7) of a solution; a pH of 7 is considered neutral. It is a measure of the hydrogen ion concentration (more specifically, the negative log of the hydrogen ion activity for glass electrodes) of a soil suspension or water.

phenology – The stage of plant development from vegetative through seed set and seed dispersal.

phenotypic plasticity – The ability of an organism to adapt itself to changing environmental conditions without any change in its genetic makeup.

phosphatic clay (slimes) – Material <0.1 mm, consisting of clay and clay-size (<2 μm) material (quartz, phosphate, etc).

photosynthesis – A process in which green plants capture the energy from the sun and use it to convert carbon dioxide and water into sugar and starch.

phytomass – All plant material in an area.

phytotoxic – Toxic to plants (e.g., inhibiting growth or survival).

pit lake – Any perennial or ephemeral water body that occupies an excavation in the land surface created for the collection of ore material. Mining activities that commonly result in pit lakes include open pit mining for precious metals, uranium, or diamonds; strip mining for coal; and quarrying for aggregate material. Pit lakes develop from either the discharge of groundwater into the excavation after dewatering wells are discontinued, the collection of surface water in the excavation, the diversion of surface water into the excavation, or a combination of these processes. With the exception of aggregate mines, pit lakes have the potential to collect and store large volumes of mining influenced water (MIW). The surface of a pit lake provides a broad interface between the terrestrial ecosystems and MIW, whereas surface water and groundwater discharge from pit lakes may impact stream ecosystems and human drinking water resources. For these reasons, regulatory agencies closely monitor the water quality of pit lakes that result from metal, coal, uranium, and diamond mining. This term does not apply to tailings ponds or storm water retention ponds unless tailings or storm water are discharged into an excavation that was originally created to collect ore material. Other names include mine lake, mine void lake (Australia), flooded strip mine (U.S. Department of Interior, Office of Surface Mining), and water-filled pit (U.S. Department of Interior, Office of Surface Mining).

placer deposit – An alluvial deposit of an economically important mineral or material, usually as a mineral gravel or sand, typically containing gold or gemstones; also a high-grade concentration of heavy mineral sands formed as lenses on present or ancient beach berms by wave action.

plant community – A group of plants occupying a common environment.

plant succession – The gradual change in plant communities in an area following a disturbance or the creation of a new substrate.

plow-in – An operation in which a large plow, usually pulled by a bulldozer, is used to open up the soil, while pipe is fed through the plow-shoe and placed at the bottom depth of the plow-shoe and the soil falls back into place behind the plow-shoe.

PLS – See *pure live seed.*

point source – (1) A discharge from a single point. (2) Any discernible, confined, and discrete conveyance, including but not limited to any pipe, ditch, channel, tunnel, conduit, well, discrete fissure, container, rolling stock, concentrated animal feeding operation, or vessel or other floating craft, from which pollutants are or may be discharged. *(U.S. Federal Water Pollution Control Act [Sec. 502(14)])*

polders – A tract of land reclaimed from the sea by diking the area and pumping out the water or by other means.

pollutant – Any organic substance, inorganic substance, a combination of organic and inorganic substances, a pathogenic organism, or heat that, when introduced into the environment, adversely impacts the usefulness of a resource.

polychaetes – Any aquatic annelid of the order *Pellucida* including both motile and sedentary forms.

pore water – Water occupying the voids in soil or sediment.

post-active acid sulfate soil – Soil that formerly contained sulfide minerals but that no longer does because the minerals have completely oxidized to form sulfates. Sulfides may still be present deeper in the soil–geologic column.

potential acidity – The acidity that could result from the oxidation of pyrite (FeS_2), sulfur, or other metal sulfides; same as *residual acidity.*

potential acid sulfate soil – Soil that contains sulfide minerals in near-surface horizons and layers that are not undergoing oxidation but which could undergo oxidation and cause extreme soil acidity if the soil is exposed to oxidizing conditions by drainage or other forms of land disturbance.

prairie protector – A normal grader blade that has three variable-height, flexible cutting edges on the bottom so that it can adapt to surface soil microtopography. It is used to remove soil from rangeland sod and place it back into the pipe ditch.

Precambrian Shield – Sometimes called the Canadian or Laurentian Shield, a massive area of glaciated bedrocks more than 570 million years old, forming a crescent from Labrador through Québec, Ontario, Manitoba, and northern Saskatchewan to the Northwest Territories of Canada.

precision – The degree of agreement among repeated measurements of the same characteristic and monitored by multiple analyses of many sample replicates and standards. It may be determined calculating the variance, standard deviation, average deviation, percent relative standard deviation, percent average deviation, or relative percent difference among samples taken from the same place at the same time, from replicate samples taken from each stage of sample preparation, or through replicate analysis of the same laboratory sample.

preparatory crop – Plowing followed by planting a residue-producing crop during the growing season before seeding perennial forages and direct seeding into the residue without further seedbed preparation.

prescribed burning – The use of fire as a management tool under specified conditions for burning a predetermined area.

press drill – An implement for seeding seed that has controls for depth of planting.

primary species – Short-lived grasses or dicots that are used to quickly control erosion; also known as nurse or companion species. These species will germinate and establish quickly.

primary succession – Development of a plant community on an area not previously occupied, such as newly exposed rock, sand, overburden, parent material, or a lava flow.

prime land – Lands that have historically been used for cropland, as designated by the U.S. Department of Agriculture; usually classes 1 and 2.

production – The total amount of mass produced by a plant, mine, aquifer, and so forth.

productivity – A measurement of soil fertility or suitability for a particular plant or crop, based on the biomass that a soil produces in a given amount of time.

propagule – Any part of a plant (e.g., a seed) that can form a new individual when separated from the parent plant.

protore – An exploration and mining industry term for mineralization that is barely subeconomic and which could become ore if the commodity price increases enough or if processing methods become more efficient, in order to make extraction and/or processing of the protore economically beneficial to the mining company. Protore mined as part of ore development and extraction is typically stockpiled separate from ore and waste piles.

pseudo-karst hydrology – The water table contained in flows through randomly oriented, highly permeable channels that are interconnected throughout the backfill.

pure live seed (PLS) – The amount of viable seed available in any given seed lot.

pyrite – An iron sulfide (FeS_2) which may form acid drainage upon exposure to oxidizing conditions and in the absence of calcium carbonate ($CaCO_3$) and other neutralizing minerals; sometimes called *fool's gold.*

QA/QC – See *quality assurance/quality control.*

quality assurance/quality control (QA/QC) – A system of procedures, checks, audits, and corrective actions to ensure that all research design and performance, environmental monitoring and sampling, and other technical and reporting activities are of the quality that meet the testing objectives.

quarry – Any open or surface workings, usually for the extraction of sand and gravel, building stone, slate, or limestone.

quartz – A very hard, inert mineral of silicon dioxide (SiO_2), commonly found in sand and sedimentary, igneous, and metamorphic rocks.

random sample – A subset of a statistical population in which each item has an equal and independent chance of being chosen.

rangeland – Any land supporting native vegetation suitable for grazing and managed by ecological principles.

range of influence (ROI) – Distance within which samples will be spatially correlated, presented in a semivariogram.

receiving waters – A river, lake, ocean, stream, or other watercourse into which wastewater or treated effluent is discharged.

receptor – An ecological entity exposed to a stressor.

reclaim – Restoring mined or disturbed land to conditions acceptable under regulatory requirements and returning the site to a safe and useful condition (e.g., grazing, recreation, agriculture, or wildlife habitat). This also refers to water that is recycled to the metallurgical process from the tailings decant water.

reclamation – Rehabilitation or return of disturbed land to productive uses; includes all activities of spoil movement, grading, and seeding; and the return of productivity equal to or exceeding that prior to its being disturbed.

reclamation material – In situ surface soil materials that are chemically and physically suitable for plant growth; also referred to as *replaced soil material* or *cover soil.* In agricultural terms these materials are generally referred to as *topsoil.*

reconstructed soil – Soil that is constructed by replacing soil materials salvaged prior to disturbance (e.g., mining) and placed on the final spoil surface.

recontouring – Shaping mined areas after backfilling.

reduced zone – The part of the soil–geologic column that contains sulfide or other reduced minerals, which commonly lies below the oxidized zone.

reference areas – Used for comparison of reclamation success and typically unmined.

refining – The purification of a crude metal product; normally the stage following smelting.

regenerative capability – The ability of a plant community to reproduce itself over time.

regolith – Soil or unconsolidated materials.

regrade – Shaping mined and processing areas after recontouring.

reinforcement seeding – Special seeding into an understocked plant community to increase species diversity, density, or utility of that community.

relational database – An electronic database comprising multiple files of related information, usually stored in tables of rows (records) and columns (fields), and allowing a link to be established between separate files that have a matching field, as a column of invoice numbers, so that the two files can be queried simultaneously by the user.

relict – An organism or group of organisms that is a surviving remnant of a formerly more widespread population.

remediation – Cleanup or other methods used to remove or contain a toxic spill or hazardous materials from a site. It is the process of correcting, counteracting, or removing an environmental problem and often refers to the removal of potentially toxic materials from soil or water by use of bacteria (bioremediation) or plants (phytoremediation).

remining – The return to underground or surface mines or previously mined areas for further ore removal by surface mining and reclaiming to current reclamation standards. It also refers to the process of mining for processing of mine and mill wastes (processed or unprocessed) to extract additional metals or other commodities due to a change in extraction technology or economics that make such remining profitable.

replaced soil material – See *reclamation material.*

representative sample – A portion of material or water that is as nearly identical in content and consistency as possible to that in the larger body of material or water being sampled.

residual acidity – Acid-producing potential of a spoil, usually caused by the presence of pyrite (FeS_2) in the spoil; same as *potential acidity.*

respiration – The metabolic oxidation of organic compounds by living organisms to produce energy.

restoration – Restoring disturbed land to the conditions that existed at the site before the disturbance occurred.

retreat – A form of high-extraction mining that involves removal of a portion or all of the ore pillars in a mine panel to increase the yield of ore from the mine. If the extraction ratio is high enough, subsidence at the surface results.

Rhizobium – Bacteria that form a symbiotic association with legumes and produce nitrogen-fixing nodules.

rhizome – A horizontal underground stem with buds. This organ serves the plant as a storage organ and a means of vegetative propagation.

riparian – Plants living or located on the bank of a natural or modified perennial waterway; may also refer to the land bordering a stream channel.

ripe – Soil materials of low n value (typically $n < 0.7$) that have relatively high load-bearing capacity.

ripping – The practice of pulling long steel teeth, usually more than 45 cm long, through soil, subsoil ore, waste rock, tailings, or overburden to break up compacted layers.

risk – A measure of the probability that damage to life, health, property, and/or the environment will occur as a result of a given hazard.

risk assessment – A qualitative and quantitative evaluation of the risk posed to human health and/or the environment by the actual or potential presence and/or use of specific pollutants. Risk assessments are conducted for a number of reasons, including to establish whether an ecological risk exists, to identify the need for additional data collection, to focus on the dangers of a specific pollutant or the risks posed to a specific site, and to help develop contingency plans and other responses to pollutant releases. Risk assessments are an important part of the U.S. Environmental Protection Agency's Superfund program and play a key role in the development and implementation of new environmental regulations.

risk characterization – The last phase of the risk assessment process that estimates the potential for adverse health or ecological effects that would occur from exposure to a stressor and evaluates the uncertainty involved.

risk communication – The exchange of information about health or environmental risks among risk assessors and managers, the general public, news media, interest groups, and so forth.

risk estimate – A description of the probability that organisms exposed to a specific dose of a chemical or other pollutant will develop an adverse response (e.g., cancer).

risk factor – A characteristic or variable associated with increased probability of a toxic effect.

risk for nonendangered species – The risk to species if anticipated toxicity is equal to or greater than LC_{50} (50% of the lethal concentration).

risk management – The process of evaluating and selecting alternative regulatory and non-regulatory responses to risk. The selection process necessarily requires the consideration of legal, economic, and behavioral factors.

rock fragment – An unattached piece of rock >2 mm in diameter that is strongly cemented or relatively resistant to rupture.

rock pile – See *waste rock.*

ROI – See *range of influence.*

room-and-pillar – A form of underground mining in which typically more than half of the ore is left in the mine as pillars to support the roof. Room-and-pillar mines are generally not expected to subside, except where retreat mining is practiced. The mining method is used for thick and/or flat-lying industrial, metal, and nonmetal mineral deposits, such as limestone, trona, and salt. In Europe, it is sometimes called *board-and-pillar.*

rotary tiller – Rotary blades used for tearing up sod and smoothing rough plowed land surfaces.

rough grading – A slope-grading technique in which the surface is left undulating, rocks are exposed, and soil clods are left intact in order to increase infiltration, reduce erosion, and create seed germination sites.

roving – Products that are extruded continuous strands of fiberglass or polypropylene used to tack down mulch.

safe site – A microenvironment specifically meeting the germination requirements of a plant species.

salic horizon – In soil taxonomy, a diagnostic horizon that has an accumulation of salts more soluble than gypsum, defined in terms of having an electrical conductivity >30 decisiemens per meter (dS/m) in a 1:1 soil/water extract over a thickness of 15 cm.

salidic sulfaquepts – In soil taxonomy, a subgroup of wet inceptisols with a sulfuric horizon within 50 cm of the soil surface that also have a salic horizon within 75 cm of the surface.

sample – A representative portion of a population.

sands – Ore or tailings particles of a size (generally >0.05 mm) and weight that readily settle in water.

SAPS – See *successive alkalinity-producing systems.*

SAR – See *sodium adsorption ratio.*

scrubber – Equipment that entraps and removes potential pollutants, using water, before they are released to the atmosphere.

scrubby – Vegetation dominated by densely growing, low, often stunted bushes or trees.

seasonal variety – A group of plants that develop and produce seed during the same general season of a growing year.

secondary species – Perennial grasses and dicots that require moderate to high soil amendments or chemical input after establishment to remain persistent.

secondary succession – The plant community development on an area previously supporting a plant community, such as abandoned cropland, cut or burned forest, and overgrazed or burned rangeland.

sedimentation – The process of depositing entrained soil particles or geologic materials from water. In a mining context, it usually results from erosion of disturbed land and is considered a negative impact to streams and other water bodies.

seed bank – Viable seeds from all types of vegetation contained in the surface soil layer.

seed bed – The soil or forest floor on which seeds fall or are placed and, if conditions are right, germinate or grow.

seed rain – The seeds that fall onto the soil as the result of dispersal, usually expressed in numbers of seeds per unit area per unit time.

semi-barren – Used in reference to vegetation, describing widely spaced, stunted trees with an abnormal amount of bare ground between them.

seminal roots – Roots that differentiate in the embryo.

semivariogram – A graphical plot of variance of an ore-grade, biological, chemical, or physical parameter as a function of distance between sampled locations.

seral – A group of plant species indicative of a certain stage of ecological succession.

sere – A series of plant communities that occupy a site over time from the initial pioneer plants to a climax community.

settlement sinking – The movement of fill material. In contrast, *differential settlement* refers to uneven, often unpredictable, movement which puts twisting stresses on buildings, pipes, and pavements and is more destructive.

sewage sludge – (1) The mainly organic, solid residual materials resulting from the treatment of sewage, often used as a soil amendment. (2) The solid, semisolid, or liquid residue that is generated during the treatment of domestic sewage in a treatment works; excludes ash that is generated during the firing of sewage sludge in a sewage sludge incinerator or grit and screenings that are generated during preliminary treatment of domestic sewage in a treatment works. (*Arizona Department of Environmental Quality*)

shale – A thinly bedded or fissile sedimentary rock formed from clay or silt particles.

shovel – Machine used to excavate ore or other minerals and to load these minerals for transport. Its bucket is loaded from the top, and the bottom is opened for emptying the contents.

shrubland – An area of land dominated by shrub species.

side slope – See *valley-side hillslope*.

silicate ore – An ore in which the valuable metal is combined with silica rather than sulfur.

sill – Sum of the nugget semivariance and semivariance due to spatial dependence in the dataset; presented in a semivariogram. Sill is also an underground hard-rock mining term (vein deposits) for a pillar separating underground levels.

single ditching – One pass or lift of the ditching equipment that removes all of the pipe ditch (trench) material; the topsoil and subsoil are mixed and removed together as one.

sintering – The use of heat to fuse ores or concentrates preparatory to further processing.

site index – Height of dominant and co-dominant canopy trees at a selected age. Age 25 is often used for pine plantations in the South. Age 50 is used for Appalachian and midwestern hardwoods.

site quality – The total effect of climate, geology, topography, and soils on the ability of a site to produce biomass.

slate – A metamorphic rock derived from shale that has been subjected to heat and/or pressure.

slimes – Material, silt or clay in size, resulting from the washing, concentration, or treatment of ground ore.

slope – The degree to which the ground angle deviates from horizontal, expressed as a percent rise over run or as a degree angle.

sludge cake – Dewatered biosolids.

slurry – Any mixture of solids and fluids that behave as a fluid and can be transported hydraulically (e.g., by pipeline). See also *tailings*.

slurry pond – An impoundment or basin designed and constructed for the dewatering and disposal of ground mineral-processing wastes.

SMCRA – Surface Mine Control and Reclamation Act of 1977

smelter – An industrial plant or process that extracts a metal from an ore or concentrate at high temperature by chemical and physical processes that occur in the molten state.

smelting – The chemical reduction of a metal from its ore or concentrate by a process usually involving fusion, so that earthy and other impurities separate as lighter and more fusible slags and can readily be removed from the reduced metal. The process commonly involves addition of reagents (fluxes) that facilitate chemical reactions and the separation of metals from impurities.

smokestack – A chimney for the disposal of waste gases and vapors into the atmosphere.

sodic soil – Sodic soils have more than 15% of their cation exchange sites occupied by sodium ions.

sodium adsorption ratio (SAR) – The relative percentage of water-soluble sodium ions compared to water-soluble calcium and magnesium ions in a soil water extract.

soil acidity (reaction or pH) classes – Descriptive terms used for ranges in pH as follows:

ultra acid: <3.5
extremely acid: 3.5–4.4
very strongly acid: 4.5–5.0
strongly acid: 5.1–5.5
moderately acid: 5.6–6.0
slightly acid: 6.1–6.5
neutral: 6.6–7.3
slightly alkaline: 7.4–7.8
moderately alkaline: 7.9–8.4
strongly alkaline: 8.5–9.0
very strongly alkaline: >9.0

soil–geologic column – Soil plus underlying geologic layers taken together.

soil microorganisms – Microorganisms such as bacteria and fungi for which soil is the natural habitat and which carry out critical biochemical processes.

soil morphology – The physical constitution of a soil profile as exhibited by the kinds, thickness, and arrangement of horizons in the profile; and by the color, texture, structure, consistence, and porosity of each horizon.

soil pH – The negative logarithm of the hydrogen ion activity of a soil suspension at a glass electrode; pH 7 indicates neutrality.

soil reconstruction – The process of reclamation of prime land in which the A, B, and/or C horizons are replaced in their relative position on some part of the disturbed mining area.

soil respiration – The sum of the respiration of all of the organisms present in soil.

soil sloughing – Soil on slopes that loses cohesion and falls to the bottom of the slope in planar or rotational failures or slumps.

soil stabilizers – Chemicals that, when applied to a surface, reduce soil movement.

soil substitute – A soil substrate derived from waste rock, overburden materials, or processing by-products.

spatial diversity – Diversity of patch size, shape, and structure within a landscape.

species diversity – A measure of the number of plant species and the distribution of individuals among species present in any given plant community.

specific conductance (Cv) – The electric current transmitted through a unit cell of water (expressed as dS/cm at 25°C).

SPGM – See *suitable plant growth material.*

spirals – Equipment used to gravity-separate the valuable heavy minerals from the waste materials. In the mineral sands industry, as the sand slurry passes down the spiral, the lighter silica sand moves to the outer side of the spiral and the heavy mineral is split off on the inner side in the process known as wet gravity separation. Trays, tables, jigs, and cones can be used also for gravity separations.

spoil – All overburden material removed above mineral resources in the process of mining; also spent ore from metallurgical leaching operations.

spoil material – Geologic material between the soil and the desired minerals that has been disturbed and not used in the reclamation process.

spolent – A proposed suborder in the U.S. system of soil taxonomy for human-influenced soils.

spontaneous colonization – The arrival of new plants at a treated disturbed site without having been deliberately introduced.

spur-end hillslope – A three-dimensional hillslope surface, curvilinear in planform and mostly convex in profile; also known as a *nose slope.*

stack – A short form for *smokestack.*

stair-step grading – A slope-grading technique that results in a slope that looks like stair steps. The riser and runners are cut into the slope so that soil erosion is minimal and moisture is retained. The runner is shallow (1–5 m deep).

stocking – In forestry, this refers to a tree density or basal area sufficient to meet certain management objectives.

stop log – A water level control structure constructed with removable boards to lower water levels in 15-to-40-cm increments.

strand lines – Refers to concentrations of heavy mineral sands deposited on former shorelines now elevated above the present water level.

stratified sample – A stratified sample is obtained by taking samples from each stratum or subgroup of a population. When sampling a population with several strata, it is generally required that the proportion of each stratum in the sample should be the same as in the population.

Stratified sampling techniques are generally used when the population is heterogeneous, or dissimilar, and where certain homogeneous, or similar, subpopulations can be isolated (strata). Simple random sampling is most appropriate when the entire population from which the sample is taken has low heterogeneity. Some reasons for using stratified sampling over simple random sampling are the cost per observation in the survey may be reduced; estimates of the population parameters may be wanted for each subpopulation; or increased accuracy at a given cost is desired. (*Statistics Glossary*)

stream channel – A trough in the landscape that conveys water and sediment; the channel formed is the product of the flow. It includes ephemeral, intermittent, and perennial stream channels. It is also known as gully, wash, run, creek, brook, and river, with the term used often dependent on the size of the channel or waterway.

stressed land – A term used to describe land rendered barren or semi-barren through soil toxicity by the long-term effects of smelter emissions. It may also describe land that has limited water-holding capacity, either as a result of compaction or other factors that cause plants to exhibit wilting while adjacent areas appear normal.

stripping ratio – The relative amount of soil or waste rock that must be removed in a surface mine to gain access to an amount of ore or mineral, expressed as a ratio of the volume of the soil and waste to ore.

structure (human) – Buildings or other types of construction at the surface above a mine; may be subject to damage from subsidence or other mining activities.

sublethal – An amount of a toxic substance that does not cause the immediate death of an organism under study in a given period of time and may or may not result in physical or physiological damage. Repeated (chronic or random) sublethal doses can cause cumulative nonlethal (chronic) or lethal effects.

subore – The economic status of mineralization in which the material may need to be mined for technical reasons but is not of sufficient quality to process for a commodity and is separately stockpiled (similar to protore). An increase in the commodity price or improved extraction technology can convert subore into ore.

subsidence – The caving in of the land surface due to the collapse of voids below the surface, often seen with near-surface or high-extraction underground mining. This can be caused by the collapse or settlement of soil materials (e.g., when water is removed from soil materials of high *n* value) or when materials high in organic matter settle from decomposition or other processes in a landfill. Subsidence can also be caused by groundwater removal from deep geologic units and subsequent compaction of those units, such as in basin-fill areas of the western United States.

subsoil – The layer or horizon (B) of the soil below the surficial topsoil or below the plow layer if the soil was cultivated in the traditional fashion.

succession – The sequence by which one set of plants replaces another over time while reacting to the environment.

successive alkalinity-producing system (SAPS) – A passive treatment system that employs a drainage system at the base of a pond, overlain by limestone and topped by a layer of organic matter. Ponded water forces water movement through organic material to strip oxygen and Fe(III), then the water flows into the limestone for generation of carbonate alkalinity.

suitable plant growth material (SPGM) – Soil material with chemical and physical properties suitable for plant establishment and growth. This may be topsoil, subsoil, or good quality spoil material.

sulfidic materials – In soil taxonomy, a diagnostic term to refer to soil materials that have the potential to become extremely acid when exposed to aerobic conditions.

sulfuricization – The process by which sulfide-bearing materials are oxidized, minerals are weathered by the sulfuric acid produced, and new mineral phases are formed from the dissolution products.

summer fallow – An unplanted field so as to allow the accumulation of soil moisture during the summer prior to fall planting.

surcharging – Inducing and speeding settlement by adding additional weight to the top of the fill.

surface mining (strip mining) – A procedure of mining that entails the complete removal of overburden material; may generally refer to either an area and/or a contour mine.

Surface Mining Control and Reclamation Act (SMCRA) – A law passed by U.S. Congress in 1977, which provides national standards for pre-mine planning and permitting of underground and surface coal mines and gives requirements for mining and reclamation practices. It also established the Office of Surface Mining Reclamation and Enforcement and authorized it to enforce rules. The law also provides for the generation of funds to reclaim abandoned metal and coal mine lands in states that contribute to the severance tax.

surface water – Water at or near the land surface, such as lakes and streams, as opposed to groundwater.

sustained yield – Continued production of usable forage over time.

swamp – A forested wetland with little peat development.

sward – An area densely covered by grass.

symbiotic nitrogen fixation – A mutually beneficial relationship between a group of bacteria (*Rhizobia*) and legumes in which a nodule is formed on the roots and atmospheric nitrogen is converted into a plant-available form.

synecology – A subdivision of ecology that studies groups of organisms associated as a unit.

tackifiers – Chemicals, often petroleum based, that are sprayed on top of mulches to keep the mulch stationary.

tailings – The solid waste product (gangue and other material) resulting from the milling and mineral concentration process (washing, concentration, and/or treatment) applied to ground ore. This term is usually used for sand to clay-sized refuse that is considered too low in mineral values to be treated further, as opposed to the concentrates that contain the valuable metals.

tailings basin – A tailings impoundment generally constructed by damming a broad, shallow valley or basin. Tailings basins are generally constructed to elevations controlled by local topography, but may be, and often are, constructed to higher elevations than surrounding topography.

tailings dam – See *tailings impoundment*.

tailings dike – Any wall or berm constructed of tailings for the purpose of retaining tailings slurries; may refer to perimeter walls on the outside crest of tailings impoundments or internal walls for the establishment of smaller cells or ponds within the larger tailings impoundment.

tailings embankment – A tailings impoundment, part or all of which is constructed above the elevation of the surrounding topography; it may be a free-standing structure and is often referred to as a tailings stack. It may also refer to any outside face of a tailings impoundment.

tailings impoundment – Any structure designed and constructed for the purpose of capturing and retaining liquid–solid slurries of mill tailings in which the solids settle. The liquid may or may not be discharged or captured for recycling after the solids have settled out of suspension. *Tailings pond* and *tailings dam* are often used interchangeably with *tailings impoundment* and *tailings storage facilities*.

tailings pond – See *tailings impoundment*.

tailings sand – The coarse fraction of the tailings stream consisting primarily of sand particles. In the case of oil sands, these tailings may contain a small amount of unrecovered bitumen.

taproot – A large primary root extending deep into the soil, bearing much smaller lateral branches.

taxa – Taxonomic groups.

thematic maps – Maps that illustrate information about a particular phenomenon or theme (e.g., U.S. Department of Agriculture soil survey maps, highway maps, topographic maps, and zoning maps).

thousand board feet (MBF) – A unit of measure to estimate usable wood-product volumes of tree trunks.

titratable acidity – The total acidity of a soil, measured by determining the quantity of a strong base required to raise soil pH to a predetermined level, often 7.0 or 8.2.

TMDL – See *total maximum daily load*.

topdressing – Topsoil substitute soil materials; may also refer to applying fertilizers to established vegetation.

topology – Topographical study of a given place in relation to its history and spatial relationships.

topsoil – The surface layer or horizon (A) of soil, generally darkened by organic matter and subject to plowing or tilling in agriculture, or its equivalent in uncultivated soils.

total maximum daily load (TMDL) – An estimate of the concentrations resulting from the total loadings (volumes multiplied by concentrations) from all sources to a water body. The TMDL is used to determine the allowable loads and provides the basis for prioritizing, establishing, or modifying controls on potential pollutant sources.

toxic material – A sample that has been tested and deemed to generate toxic concentrations of pollutants when associated with mining or processing. Other materials may contain toxic levels of various elements to plants and/or animals.

toxicant – A harmful substance or agent that may injure an exposed organism.

toxicity – A property of a substance that indicates its ability to cause physical and/or physiological harm to an organism (plant, animal, or human), usually under particular conditions and above a certain concentration limit, below which no toxicity effects have been observed.

treat and/or cover – The reclamation practice for slurry requiring limestone amendment and/or soil cover.

trigger-factor effect – The effect that occurs when a change in a single environmental factor or event, such as an earthquake or flood, causes an irreversible sequence of events.

triple ditching – Three passes or lifts of the ditching equipment used to remove the pipe ditch (trench) material. Usually topsoil is removed in one pass (first lift), upper subsoil (mid-soil) is removed with the second pass (second lift), and the lower subsoil is removed with the third pass (third lift). This process results in three separate soil stockpiles on the pipeline right-of-way.

TSF – tailings storage facillity. See also *tailings impoundment.*

tubellarians – Any aquatic free-swimming *Platyhelminth* of the class *Tubellaria.*

tunnel – See *adit.*

ultisol – In soil taxonomy, the order of mineral soils that have horizons of clay accumulation and a low base saturation (<35%) deep in the soil profile.

understory – In a forest or shrubland, the layer or layers of vegetation below the main canopy.

unoxidized zone – See *reduced zone.*

unripe – Soil materials with high *n* value (typically >0.7) that are soft and have low or poor load-bearing capacity.

USBM – U.S. Bureau of Mines

USDA – U.S. Department of Agriculture

USGS – U.S. Geological Survey

valley-head hillslope – A three-dimensional hillslope surface, curvilinear in planform and mostly concave in profile; also known as a *hollow.*

valley-side hillslope – A three-dimensional hillslope surface, straight in planform and convex; and straight, concave, or complex in profile; also known as a *side slope* or *valley-side.*

vegetative reproduction – Reproduction in which the propagule is other than a spore or a seed (e.g., a bulb, rhizome, or corm).

ventilation drift or shaft – A horizontal adit or tunnel or vertical shaft in a mine with the prime purpose of exchanging gases with the outside atmosphere.

vigor – A qualitative measure of plant health.

volcanic ash – A usually fine-grained rock, formed by volcanic action at the earth's surface, consisting of rock, mineral, and volcanic glass fragments smaller than 2 mm (0.1 in.) in diameter; ash particles less than 0.025 mm (1/1,000th of an inch) in diameter are common. Ash is extremely abrasive, mildly corrosive, and electrically conductive, especially when wet. Volcanic ash is created during explosive eruptions by the shattering of solid rocks and violent separation of magma (molten rock) into tiny particles. (*http://volcanoes.usgs.gov/products/pglossary/ash.html*)

volunteer species – The plant species that naturally invade a site without human assistance.

warm-season plant – Plants that make most of their growth during the warmer part of the growing season; a C4 plant.

waste rock – Barren or mineralized rock that has been mined but is of insufficient value to warrant treatment and, therefore, is removed ahead of the metallurgical processes and disposed of on site. The term is usually used for wastes that are larger than sand-sized material and can be up to large boulders in size; also referred to as *waste rock dump* or *rock pile*.

water balance – An accounting of the inflow to, outflow from, and storage changes of water in a hydrologic unit over a fixed period.

water body – Any natural or artificial stream, river, or drainage with perceptible flow at the time of crossing during pipeline construction; and other permanent to ephemeral water bodies such as ponds and lakes.

water cycle – The process by which water travels in a sequence from the air (condensation) to the earth (precipitation) and returns to the atmosphere (evaporation).

water quality standards – State-adopted and EPA-approved ambient standards for water bodies. The standards prescribe the use of the water body and establish the water quality criteria that must be met to protect designated uses.

water use efficiency (WUE) – The amount of grain or forage produced per water volume used.

watershed – The land area that drains into a stream. The watershed for a major river may encompass a number of smaller watersheds that ultimately combine at a common point.

weathering – Process whereby earthy or rocky materials are changed in color, texture, composition, or form (with little or no transportation) by exposure to atmospheric agents.

weed – A plant that aggressively colonizes habitats where it is not wanted.

wet mill – See *concentrator*.

wetlands – Land areas containing ponded water or saturated surface soil for some portion of the growing season. Those with standing water for long periods may be mined only under special conditions, and the owner usually must reconstruct more acres of wetlands than originally disturbed.

wheatland plow – A single gang of upright discs spaced along one axle and used on previously cultivated land for light tillage.

woodland – Land area that supports primarily woody shrubs and subcommercial trees.

woody draw – Areas of woody plant communities along waterways or intermittent streams and drainages where favorable soil water occurs; may also be referred to as a *hardwood draw*.

workings – The entire system of openings (underground as well as at the surface) in a mine.

WUE – See *water use efficiency*.

xeric – Dry, with limited moisture or water.

zero point of charge – The condition of a mineral sample where the negative and positive charges are equal and the net charge is zero.

zone of weathering – See *oxidized zone*.

USEFUL WEB SITES

General geological dictionary – www.webref.org/geology/geology.htm

Glossary of general mining and processing terms – www.northernminer.com/tools/glossary.asp

Mineral Information Finder, a very comprehensive Web site – www.rocksandminerals.com/glossary.htm

Minerals Information Institute, photos and descriptions of minerals – www.mii.org/mineralphotos.html

19th and 20th Century mining terms, based in Australia – www.heritagearchaeology.com.au/publications/glossary.htm

SOURCES FOR DEFINITIONS

American Society of Agronomy (www.agronomy.org). Entries can be found in an unedited form in Barnhisel et al.

Arizona Department of Environmental Quality: www.azdeq.gov/environ/water/permits/programs.html#sewer

American Society for Metals Glossary: Definitions relating to metals and metalworking, in Metals handbook, 8th ed. Vol. 1, Properties and selection of metals. Metals Park, Ohio: American Society for Metals, 1961. pp. 1–41.

ASTM E1915: *Annual Book of ASTM Standards*, Vol. 03.06. West Conshohocken, PA: American Society for Testing and Materials, 2007.

ASTM STP No. 148-D: *Manual on Industrial Water and Industrial Waste Water*, Appendix II, Glossary. Special Technical Publication 148–D, 2nd ed. West Conshohocken, PA: American Society for Testing and Materials, 1959. pp. 635–643.

Barnhisel, R.I., R.G. Darmody, and W.L. Daniels, eds., *Reclamation of Drastically Disturbed Lands*. Published by the American Society of Agronomy, Crop Science Society of America, and Soil Science Society of America. Agronomy Monograph 41, 2000.

Nelson, A., *Dictionary of Mining*. New York: Philosophical Library, Inc., 1965.

Pryor, E.J., *Dictionary of Mineral Technology.* London: Mining Publications, Ltd., 1963.

Smith, K.S., and Huyck, H.L.O., An overview of the abundance, relative mobility, bioavailability, and human toxicity of metals, Chapter 2, in Plumlee, G.S., and Logsdon, M.J., eds., *The Environmental Geochemistry of Mineral Deposits, Part A: Processes, Techniques, and Health Issues,* Reviews in Economic Geology, Vol. 6A. Littleton, CO: Society of Economic Geologists, 1999, pp. 29–70.

Statistics Glossary: www.stats.gla.ac.uk/steps/glossary/index.html

Thrush, P.W., *A Dictionary of Mining, Mineral, and Related Terms,* compiled and edited by Paul W. Thrush and the Staff of the Bureau of Mines. Washington, DC: U.S. Bureau of Mines, 1968.

U.S. Bureau of Mines Federal Mine Safety Code—Bituminous Coal and Lignite Mines, Pt. I Underground Mines, October 8, 1953.

U.S. Geological Survey: http://volcanoes.usgs.gov/products/pglossary/ash.html

References

Allgaier, F.K. 1997. Environmental effects of mining. In *Mining Environmental Handbook: Effects of Mining on the Environment and American Controls on Mining*. Edited by J.J. Marcus. London: Imperial College Press. pp. 132–189.

Alpers, C.N., Blowes, D.W., Nordstrom, D.K., and Jambor, J.L. 1994. Secondary minerals and acid mine-water chemistry. In *The Environmental Geochemistry of Sulfide Mine-Wastes: Short Course Handbook*, Vol. 22. Edited by D.W. Blowes and J.L. Jambor. Waterloo, ON: Mineralogical Association of Canada. pp. 245–270.

Biswas, A.K., and Davenport, W.G. 1992. *Extractive Metallurgy of Copper*, 3rd ed. Oxford: Pergamon Press.

Boening, D.W., and Chew, C.M. 1999. A critical review: General toxicity and environmental fate of three aqueous cyanide ions and associated ligands. *Water, Air, Soil Pollut.* 109(1–4):67–79.

Brown, R. 1989. Water management at Brenda Mines. In *Proceedings of the 13th Annual British Columbia Mine Reclamation Symposium*, Vernon, BC, June 7–9. Victoria, BC: Ministry of Energy, Mines and Petroleum Resources. pp. 8–17.

Busenberg, E., and Clemency, C. 1976. The dissolution kinetics of feldspars at 25°C and 1 atmosphere CO_2 partial pressure. *Geochim. Cosmochim. Acta* 40:41–49.

Busenberg, E., and Plummer, L.N. 1982. The kinetics of dissolution of dolomite in CO_2–H_2O systems at 1.5° to 65°C and 0 to 1 atm P_{CO2}. *Am. J. Sci.* 282:45–78.

———. 1986. *A Comparative Study of the Dissolution and Crystal Growth Kinetics of Calcite and Aragonite.* Bulletin 1578:139–168. U.S. Geological Survey.

Caruccio, F.T., and Geidel, G. 1980. *The Geologic Distribution of Pyrite and Calcareous Material and Its Relationship to Overburden Sampling*. Information Circular IC-8863:2–12. Washington, DC: U.S. Bureau of Mines.

Caruccio, F.T., Hossner, L.R., and Geidel, G. 1988. Pyritic materials: Acid drainage, soil acidity, and liming. In *Reclamation of Surface-Mined Lands*, Vol. 1. Boca Raton, FL: CRC Press. pp. 159–190.

Chou, L., Garrels, R.M., and Wollast, R. 1989. Comparative study of the kinetics and mechanisms of dissolution of carbonate minerals. *Chem. Geol.* 78(3–4):269–282.

Cole, K.A., and Kirkpatrick, A. 1983. Cyanide heap leaching in California: Methods used, plus health and safety aspects. *Calif. Geol.* 36(9):187–194.

Cox, D.P., and Singer, D.A., eds. 1986. *Mineral Deposit Models*. Bulletin 1693. U.S. Geological Survey.

Cravotta, C.A., III. 1994. Secondary iron-sulfate minerals as sources of sulfate and acidity. In *Environmental Geochemistry of Sulfide Oxidation*. Edited by C.N. Alpers and D.W. Blowes. ACS Symposium Series 550. Washington, DC: American Chemical Society. pp. 345–364.

Desborough, G.A., Briggs, P.H., Mazza, N., and Driscoll, R. 1998. *Acid-Neutralizing Potential of Minerals in Intrusive Rocks of the Boulder Batholith in Northern Jefferson County, Montana*. Open-File Report 98-364. Denver, CO: U.S. Geological Survey.

Dorr, J.V.N., and Bosqui, F.L. 1950. *Cyanidation and Concentration of Gold and Silver Ores*. New York: McGraw-Hill.

Drever, J.L. 1988. *The Geochemistry of Natural Waters*. Englewood Cliffs, NJ: Prentice Hall. p. 144.

Du Bray, E.A., ed. 1995. *Preliminary Compilation of Descriptive Geoenvironmental Mineral Deposit Models*. Open-File Report 95-831. U.S. Geological Survey. http://pubs.usgs.gov/of/1995/ofr-95-0831. Accessed October 2007.

Eary, L.E., and Williamson, M.A. 2006. Simulations of the neutralizing capacity of silicate rocks in acid mine drainage environments. In *Seventh International Conference on Acid Rock Drainage*. Edited by R.I. Barnhisel. Lexington, KY: American Society of Mining and Reclamation. pp. 564–577.

Eckstrand, O.R., ed. 1984. *Canadian Mineral Deposit Types: A Geological Synopsis*. Economic Geology Report 36. Ottawa, Canada: Geological Survey of Canada.

EPA (U.S. Environmental Protection Agency). 1971. *Inorganic Sulfur Oxidation by Iron-Oxidizing Bacteria*. Water Pollution Control Research Series 14010. Washington, DC: U.S. Government Printing Office.

———. 1994. *Treatment of Cyanide Heap Leaches and Tailings*. Technical Report EPA 530-R-94-037, NTIS PB94-201837. Washington, DC: U.S. EPA.

———. 1995. *Profile of the Metal Mining Industry*. EPA Office of Compliance Sector Notebook Project, EPA/310-R-95-008. www.epa.gov/compliance/resources/publications/assistance/sectors/notebooks/mining.html. Accessed October 2007.

———. 1997. *Introduction to Hard Rock Mining: A CD-ROM Application*. EPA 530-C-97-005. Washington, DC: U.S. EPA.

———. 2003. The hydrologic cycle. www.epa.gov/seahome/groundwater/src/hydrocyc.htm. Accessed April 2003.

———. 2006. Drinking water contaminants. www.epa.gov/safewater/mcl.html#mcls. Accessed May 2006.

Evangelou, V.P. 1995. *Pyrite Oxidation and Its Control*. Boca Raton, FL: CRC Press.

———. 1998. *Environmental Soil and Water Chemistry: Principles and Applications*. New York: John Wiley and Sons.

Evangelou, V.P., and Zhang, Y.L. 1995. A review: Pyrite oxidation mechanisms and acid mine drainage prevention. *Crit. Rev. Environ. Sci. Technol.* 25(2):141–199.

Farmer, I. 1999. Room-and-pillar mining. In *Mining Engineering Handbook*. Edited by H.L. Hartman. Littleton, CO: SME. pp. 1681–1701.

Fennemore, G.G., Neller, W.C., and Davis, A. 1998. Modeling pyrite oxidation in arid environments. *Environ. Sci. Technol.* 32:2680–2687.

Ficklin, W.H., Plumlee, G.S., Smith, K.S., and McHugh, J.B. 1992. Geochemical classification of mine drainages and natural drainages in mineralized areas. In *Proceedings of the Seventh International Water–Rock Interaction Conference*, Park City, UT. Edmonton, AB: International Association of Geochemistry and Cosmochemistry and Alberta Research Council, Sub-Group on Water–Rock Interaction. pp. 381–384.

Ford, W.E. 1929. *Dana's Manual of Mineralogy*. New York: John Wiley and Sons.

Garrels, R.M., and Christ, C.L. 1965. *Solutions, Minerals and Equilibria*. Boston: Jones and Bartlett.

Gertsch, R.E., and Bullock, R.L., eds. 1998. *Techniques in Underground Mining*. Littleton, CO: SME.

Gilchrist, J.D. 1989. *Extraction Metallurgy*, 3rd ed. Oxford: Pergamon Press.

Gill, C.B. 1991. *Materials Beneficiation*. New York: Springler-Verlag.

Goleva, G.A., Polyakov, V.A., and Nechayeva, T.P. 1970. Distribution and migration of lead in ground waters. *Geochem. Int.* 7:256–268.

Greenberg, J., and Tomson, M. 1992. Precipitation and dissolution kinetics and equilibria of aqueous ferrous carbonate vs. temperature. *Appl. Geochem.* 7(2):185–190.

Guilbert, J.M., and Park, C.F. 1986. *The Geology of Ore Deposits*. New York: W.H. Freeman.

Hartman, H.L. 1992. Introduction to mining. In *Mining Engineering Handbook*. Edited by H.L. Hartman. Littleton, CO: SME. pp. 3–42.

Hem, J.D. 1970. *Study and Interpretation of the Chemical Characteristics of Natural Water*. Water-Supply Paper 1473. Alexandria, VA: U.S. Geological Survey.

Hornberger, R.J., Lapakko, K.A., Krueger, G.E., Bucknam, C.H., Ziemkiewicz, P.F., van Zyl, D.J.A., and Posey, H.H. 2000. The Acid Drainage Technology Initiative (ADTI). In *ICARD 2000: Proceedings of the Fifth International Conference on Acid Rock Drainage*. Littleton, CO: SME. pp. 41–50.

IIED (International Institute for Environment and Development). 2002. *Breaking New Ground: Mining, Minerals, and Sustainable Development*. London: Earthscan. www.iied.org/mmsd/finalreport/index.html. Accessed October 2007.

Jambor, J.L. 1994. Mineralogy of sulfide-rich tailings and their oxidation products. In *The Environmental Geochemistry of Sulfide Mine-Wastes: Short Course Handbook*, Vol. 22. Edited by D.W. Blowes and J.L. Jambor. Waterloo, ON: Mineralogical Association of Canada. pp. 103–132.

———. 2002. Static tests of neutralization potentials of silicate and aluminosilicate minerals. *Environ. Geol.* 43:1–17.

Jerz, J.K., and Rimstidt, J.D. 2004. Pyrite oxidation in moist air. *Geochim. Cosmochim. Acta* 68:701–714.

Julin, D.E. 1992. Block caving. In *Mining Engineering Handbook*. Edited by H.L. Hartman. Littleton, CO: SME. pp. 1815–1836.

Kleinmann, R.L.P., ed. 2000. *Prediction of Water Quality at Surface Coal Mines*. Morgantown, WV: National Mine Land Reclamation Center.

Kleinmann, R.L.P., Crerar, D.A., and Pacelli, R.R. 1981. Biogeochemistry of acid mine drainage and a method to control acid formation. *Min. Eng.* 3(3):300–305.

Kvapil, R. 1992. Sublevel caving. In *Mining Engineering Handbook*. Edited by H.L. Hartman. Littleton, CO: SME. pp. 1789–1814.

Kwong, Y.T.J., and Ferguson, K.D. 1990. Water chemistry and mineralogy at Mt. Washington: Implications to acid generation and metal leaching. In *Acid Mine Drainage: Designing for Closure*. Edited by J.W. Gadsby, J.A. Malick, and S.J. Day. Vancouver, BC: Bitech Publishers. pp. 217–230.

Lapakko, K.A. 1993. *Field Dissolution of Test Piles of Duluth Complex Rock*. Report to the U.S. Bureau of Mines, Salt Lake City Research Center. Unpublished.

Lapakko, K.A., and Antonson, D.A. 1994. Oxidation of sulfide minerals present in Duluth Complex rock: A laboratory study. In *Environmental Geochemistry of Sulfide Oxidation*. ACS Symposium Series 550. Washington, DC: American Chemical Society. pp. 593–607.

Lapakko, K.A., Antonson, D.A., and Wagner, J.R. 1997. Mixing of limestone with finely-crushed acid-producing rock. In *Proceedings of the Fourth International Conference on Acid Rock Drainage*, Vol. 3. Vancouver, BC, May 31–June 6. pp. 1345–1360.

Lapakko, K., Haub, J., and Antonson, D. 1998. Effects of dissolution time and particle size on kinetic test results. SME Preprint 98-114. Littleton, CO: SME.

Leonard, J.W., III, and Hardinge, B.C., eds. 1991. *Coal Preparation*, 5th ed. Littleton, CO: SME.

Lewis, A. 1983. Leaching and precipitation technology for gold and silver ores. *Eng. Min. J.* (June):48–56.

Lindgren, W. 1933. *Mineral Deposits*, 4th ed. New York: McGraw-Hill.

Lindgren, W., Graton, L.C., and Gordon, C.H. 1910. *The Ore Deposits of New Mexico*. Professional Paper 68. U.S. Geological Survey.

Mackey, T.S., and Prengaman, R.D., eds. 1990. *Lead-Zinc '90*. Warrendale, PA: Minerals, Metals and Materials Society.

Marsden, J., and House, I. 2006. *The Chemistry of Gold Extraction*, 2nd ed. Littleton, CO: SME.

Mason, K. 1998. Cyanide: Myths, misconceptions and management. *AusIMM* 3:65–70.

McKibben, M.A., and Barnes, H.L. 1986. Oxidation of pyrite in low-temperature acidic solutions: Rate laws and surface textures. *Geochim. Cosmochim. Acta* 50:1509–1520.

McLemore, V.T. 2001. *Silver and Gold Resources in New Mexico*. Resource Map 21. Socorro, NM: New Mexico Bureau of Mines and Mineral Resources.

MEM (Mining Environmental Management). 2001. Tailings. *Min. Environ. Manage.* (May):40.

Meyer, J.W., and Leonardson, R.W. 1990. Tectonic, hydrothermal and geomorphic controls on alteration scar formation near Questa, New Mexico. In *Tectonic Development of the Southern Sangre de Cristo Mountains, New Mexico*. Guidebook 41. Edited by P.W. Bauer, S.G. Lucas, C.K. Mawer, and W.C. McIntosh. Socorro, NM: New Mexico Geological Society. pp. 417–422.

Mills, C. 1998a. The role of microorganisms in acid rock drainage. http://technology.infomine.com/enviromine/ard/Microorganisms/roleof.htm. Accessed October 2007.

Mills, C. 1998b. Variables and uncertainties in the geochemical characterization of mineral samples by kinetic methods. In *Summary Notes, Fifth Annual BC MEM–MEND 2000 Metal Leaching and ARD Workshop*, Vancouver, BC, December 9–10, Section C.2. Vancouver, BC: Canada Centre for Mineral and Energy Technology. pp. 1–12.

MMSD (Mining, Minerals and Sustainable Development). 2002. *Seven Questions to Sustainability: How to Assess the Contribution of Mining and Minerals Activities*. Winnipeg, MT: International Institute for Environment and Development. www.iisd.org/pdf/2002/mmsd_sevenquestions.pdf. Accessed October 2007.

Morin, K.A., and Hutt, N.M. 1994. Observed preferential depletion of neutralization potential over sulfide minerals in kinetic tests: Site specific criteria for safe NP/AP ratios. In *Proceedings of the International Land Reclamation and Mine Drainage Conference on the Abatement of Acidic Drainage*, Pittsburgh, PA, April 24–29. Washington, DC: U.S. Government Printing Office. pp. 148–156.

Mortimer, C.E. 1967. *Chemistry, a Conceptual Approach*. New York: Van Nostrand Reinhold.

Mudder, T. 2001. The cyanide guide. *Min. Environ. Manage.* 9(3):8–12.

Munroe, E.A., McLemore, V.T., and Dunbar, N.W. 2000. Mine waste rock pile geochemistry and mineralogy in southwestern New Mexico, USA. In *ICARD 2000: Proceedings of the Fifth International Conference on Acid Rock Drainage*. Littleton, CO: SME. pp. 1327–1336.

Nelson, M.B. 1978. Kinetics and mechanisms of the oxidation of ferrous sulfide. Ph.D. thesis, Stanford University, Palo Alto, CA.

Nesbitt, H.W., and Jambor, J.L. 1998. Role of mafic minerals in neutralizing ARD, demonstrated using a chemical weathering methodology. In *Modern Approaches to Ore and Environmental Mineralogy. Short Course Series*, Vol. 27. Edited by L.J. Cabri and D.J. Vaughan. Ottawa, ON: Mineralogical Association of Canada. pp. 403–421.

———. 1999. Some fundamentals of aqueous geochemistry. In *The Environmental Geochemistry of Mineral Deposits, Part A: Processes, Techniques, and Health Issues*. Edited by G.S. Plumlee and M.J. Logsdon. Reviews in Economic Geology, Vol. 6A. Littleton, CO: Society of Economic Geologists. pp. 117–123.

Nordstrom, D.K., and Alpers, C.N. 1999. Geochemistry of acid mine waters. In *The Environmental Geochemistry of Mineral Deposits, Part A: Processes, Techniques, and Health Issues*. Edited by G.S. Plumlee and M.J. Logsdon. Reviews in Economic Geology, Vol. 6A. Littleton, CO: Society of Economic Geologists. pp. 133–160.

North, R.M., and McLemore, V.T. 1986. *Silver and Gold Occurrences in New Mexico*. Resource Map 15. Socorro, NM: New Mexico Bureau of Mines and Mineral Resources.

Pentz, S.B., and Kostaschuk, R.A. 1999. Effect of placer mining on suspended sediment in reaches of sensitive fish habitat. *Environ. Geol.* 37:78–89.

Plumlee, G.S. 1999. The environmental geology of mineral deposits. In *The Environmental Geochemistry of Mineral Deposits, Part A: Processes, Techniques, and Health Issues*. Edited by G.S. Plumlee and M.J. Logsdon. Reviews in Economic Geology, Vol. 6A. Littleton, CO: Society of Economic Geologists. pp. 71–116.

Plumlee, G.S., and Nash, J.T. 1995. *Geoenvironmental Models of Mineral Deposits—Fundamentals and Application*. Open-File Report 95-831. Denver, CO: U.S. Geological Survey. pp. 1–9.

Plumlee, G.S., Smith, K.S., and Ficklin, W.H. 1994. *Geoenvironmental Models of Mineral Deposits, and Geology-Based Mineral-Environmental Assessments of Public Lands*. Open-File Report 94-203. Reston, VA: U.S. Geological Survey.

Plumlee, G.S., Smith, K.S., Montour, M.R., Ficklin, W.H., and Mosier, E.L. 1999. Geologic controls on the composition of natural waters and mine waters. In *The Environmental Geochemistry of Mineral Deposits, Part B: Case Studies and Research Topics*. Edited by L.H. Filipek and G.S. Plumlee. Reviews in Economic Geology, Vol. 6B. Littleton, CO: Society of Economic Geologists. pp. 373–432.

Posey, H.H., Renkin, M.L., and Woodling, J. 2000. Natural acid drainage in the Upper Alamosa River of Colorado. In *ICARD 2000: Proceedings of the Fifth International Conference on Acid Rock Drainage*. Littleton, CO: SME. pp. 485–498.

Pugh, C.E., Hossner, L.R., and Dixon, J.B. 1984. Oxidation rate of iron sulfides as affected by surface area, morphology, oxygen concentration, and autotrophic bacteria. *Soil Sci.* 137(5):309–314.

Ranville, J.F., and Schmiermund, R.L. 1999. General aspects of aquatic colloids in environmental geochemistry. In *The Environmental Geochemistry of Mineral Deposits, Part A: Processes, Techniques, and Health Issues*. Edited by G.S. Plumlee and M.J. Logsdon. Reviews in Economic Geology, Vol. 6A. Littleton, CO: Society of Economic Geologists. pp. 183–199.

Rimstidt, J.D., Chermak, J.A., and Gagen, P.M. 1994. Rates of reaction of galena, sphalerite, chalcopyrite, and arsenopyrite with Fe(III) in acidic solutions. In *Environmental Geochemistry of Sulfide Oxidation*. Edited by C.N. Alpers and D.W. Blowes. ACS Symposium Series 550. Washington, DC: American Chemical Society. pp. 2–13.

Ritchie, A.I.M. 1994. The waste rock environment. In *The Environmental Geochemistry of Sulfide Mine-Wastes: Short Course Handbook*, Vol. 22. Edited by D.W. Blowes and J.L. Jambor. Waterloo, ON: Mineralogical Association of Canada. pp. 131–161.

Roberts, R.G., and Sheahan, P.A., eds. 1988. Ore deposit models. Geoscience Canada, Reprint Series 3. St. John's, NL: Geological Association of Canada.

Schlitt, W.J. 1999. Surface and hybrid mining: Aqueous extraction methods. In *Mining Engineering Handbook*. Edited by H.L. Hartman. Littleton, CO: SME. pp. 1453–1517.

Schmiermund, R.L., and Drozd, M.A. 1997. Acid mine drainage and other mining-influenced waters. In *Mining Environmental Handbook*. Edited by J.J. Marcus. London: Imperial College Press. pp. 599–617.

Seal, R.R., II, and Foley, N.K., eds. 2002. *Progress on Geoenvironmental Models for Selected Mineral Deposit Types*. Open-File Report 02-195. Reston, VA: U.S. Geological Survey. http://pubs.usgs.gov/of/2002/of02-195. Accessed October 2007.

Seal, R.R., II, Wanty, R.B., and Foley, N.K. 2000. Introduction to geoenvironmental models of mineral deposits. In *Geoenvironmental Analysis of Ore Deposits*. Short course presented at the Fifth International Conference on Acid Mine Drainage, Denver, CO, May 21.

Sheahan, P.A., and Cherry, M.E., eds. 1993. *Ore Deposit Models*, Vol. II. Geoscience Canada, Reprint Series 6. St. John's, NL: Geological Association of Canada.

Sillitoe, R.H. 1972. *Relation of Metal Provinces in Western America to Subduction of Oceanic Lithosphere*. Bulletin 83. Boulder, CO: Geological Society of America. pp. 813–818.

———. 1981. Ore deposits of the Cordilleran and island arc settings. In *Relations of Tectonics to Ore Deposits in the Southern Cordillera*. Edited by W.R. Dickinson and W.D. Payne. Arizona Geological Society Digest, Vol. 14. Tucson, AZ: Arizona Geological Society. pp. 49–69.

Singer, P.C., and Stumm, W. 1970. Acid mine drainage: The rate determining step: *Science* 167:1121–1123.

Skousen, J., Rose, A., Geidel, G., Foreman, J., Evans, R., Hellier, W., and ARC (Avoidance and Remediation Committee of the Acid Drainage Technology Initiative). 1998. *Handbook of Technologies for Avoidance and Remediation of Acid Mine Drainage*. Morgantown, WV: National Mine Land Reclamation Center.

Smith, A. 1994. The geochemistry of cyanide in mill tailings. In *The Environmental Geochemistry of Sulfide Mine-Wastes: Short Course Handbook*, Vol. 22. Edited by D.W. Blowes and J.L. Jambor. Waterloo, ON: Mineralogical Association of Canada. pp. 293–332.

Smith, A., and Struhsacker, D.W. 1988. Cyanide geochemistry and detoxification regulations. In *Introduction to Evaluation, Design and Operation of Precious Metal Heap Leaching Projects*. Edited by D.J.A. van Zyl, I.P.G. Hutchison, and J.E. Kiel. Littleton, CO: SME. pp. 275–292.

Smith, A.C.S., and Mudder, T.I. 1999. The environmental geochemistry of cyanide. In *The Environmental Geochemistry of Mineral Deposits, Part A: Processes, Techniques, and Health Issues*. Edited by G.S. Plumlee and M.J. Logsdon. Reviews in Economic Geology, Vol. 6A. Littleton, CO: Society of Economic Geologists. pp. 229–248.

Smith, E.E., and Shumate, K.S. 1970. *Sulfide to Sulfate Reaction Mechanism, a Study of the Sulfide to Sulfate Reaction Mechanism As It Relates to the Formation of Acid Mine Waters*. Water Pollution Control Research Series 14010. Washington, DC: U.S. Government Printing Office.

Smith, K.S. 1999. Metal sorption on mineral surfaces: An overview with examples relating to mineral deposits. In *The Environmental Geochemistry of Mineral Deposits, Part A: Processes, Techniques, and Health Issues*. Edited by G.S. Plumlee and M.J. Logsdon. Reviews in Economic Geology, Vol. 6A. Littleton, CO: Society of Economic Geologists. pp. 161–182.

Smith, K.S., and Huyck, H.L.O. 1999. An overview of the abundance, relative mobility, bioavailability, and human toxicity of metals. In *The Environmental Geochemistry of Mineral Deposits, Part A: Processes, Techniques, and Health Issues*. Edited by G.S. Plumlee and M.J. Logsdon. Reviews in Economic Geology, Vol. 6A. Littleton, CO: Society of Economic Geologists. pp. 29–70.

Smith, K.S., Ramsey, C.A., and Hageman, P.L. 2000. Statistically-based sampling strategy for sampling the surface material of mine-waste dumps for use in screening and prioritizing historical dumps on abandoned mines lands. In *ICARD 2000: Proceedings of the Fifth International Conference on Acid Rock Drainage*. Littleton, CO: SME. pp. 1453–1461.

Smith, S.S. 2001. Statistical summary. In *Minerals Yearbook 2001*. Reston, VA: U.S. Geological Survey. http://minerals.usgs.gov/minerals/pubs/commodity/statistical_summary/871499.pdf. Accessed October 2007.

Sobek, A.A., Schuller, W.A., Freeman, J.R., and Smith, R.M. 1978. *Field and Laboratory Methods Applicable to Overburdens and Minesoils*. EPA 600/2-78-054. Washington, DC: U.S. Environmental Protection Agency.

Stout, K.S. 1967. *Mining Methods and Equipment Illustrated*. Bulletin 63. Butte: Montana Bureau of Mines and Geology.

———. 1980. *Mining Methods and Equipment: Mining Informational Services*. New York: McGraw-Hill.

Stumm, W., and Morgan, J.J. 1981. *Aquatic Chemistry—An Introduction Emphasizing Chemical Equilibria in Natural Waters*. New York: John Wiley and Sons.

Susak, N.J. 1994. Alteration factors affecting ore deposition. In *Alteration and Alteration Processes Associated with Ore-Forming Systems*. Edited by D.R. Lentz. Short Course Notes, Vol. 11. St. John's, NL: Geological Association of Canada. pp. 115–130.

Thomas, L.J. 1973. *An Introduction to Mining*. Sydney: Hicks, Smith and Sons.

Tilsley, J.E. 1990. Genetic considerations relating to some uranium ore deposits. In *Ore Deposit Models*. Edited by R.G. Roberts and P.A. Sheahan. Geoscience Canada, Reprint Series 3. St. John's, NL: Geological Association of Canada. pp. 91–102.

USBM (U.S. Bureau of Mines). 1994. *This Is Mining*. U.S. Department of the Interior. http://imcg.wr.usgs.gov/usbmak/thisis.html. Accessed January 2003.

Wagner, L.A. 2002. *Materials in the Economy—Material Flows, Scarcity, and the Environment*. Circular 1221. Denver, CO: U.S. Geological Survey. http://pubs.usgs.gov/circ/2002/c1221. Accessed October 2007.

Weight, W.D., and Sonderegger, J.L. 2001. *Manual of Applied Field Hydrology*. New York: McGraw-Hill.

White, W.W., III, and Jeffers, T.H. 1994. Chemical predictive modeling of acid mine drainage from metallic sulfide-bearing rock. In *Environmental Geochemistry of Sulfide Oxidation*. ACS Symposium Series 550. Washington, DC: American Chemical Society. pp. 608–630.

White, W.W., III, Lapakko, K.A., and Cox, R.L. 1999. Static-test methods most commonly used to predict acid-mine drainage: Practical guidelines for use and interpretation. In *The Environmental Geochemistry of Mineral Deposits*. Edited by G.S. Plumlee and M.J. Logsdon. Reviews in Economic Geology, Vol. 6A. Littleton, CO: Society of Economic Geologists. pp. 325–338.

Whitehouse, A.E., Posey, H.H., Gillis, T.D., Long, M.B., and Mulyana, A.A.S. 2006. Mercury discharges from small scale gold mines in North Sulawesi, Indonesia: Managing a change from mercury to cyanide. Presented at the Seventh ICARD Meeting, St. Louis, MO, March 26. Lexington, KY: American Society of Mining and Reclamation. pp. 2354–2368.

Whyte, J., and Danielson, V., eds. 1998. *Mining Explained: A Layman's Guide*. Don Mills, ON: Northern Miner.

Williams-Jones, A.E., Wood, S.A., Mountain, B.W., and Gammons, C.H. 1994. Experimental water–rock interaction: Applications to ore-forming hydrothermal systems. In *Alteration and Alteration Processes Associated with Ore-Forming Systems*. Edited by D.R. Lentz. Short Course Notes, Vol. 11. St. John's, NL: Geological Association of Canada. pp. 131–160.

Williamson, M.A., and Rimstidt, J.D. 1994. The kinetics and electrochemical rate-determining step of aqueous pyrite oxidation. *Geochim. Cosmochim. Acta* 58:5443–5454.

Williamson, M.A., Kirby, C.S., and Rimstidt, J.D. 2006. Iron dynamics in acid mine drainage. Presented at the Seventh ICARD Meeting, St. Louis, MO, March 26. Lexington, KY: American Society of Mining and Reclamation. pp. 2411–2423.

Wills, B.A. 1992. *Mineral Processing Technology*, 5th ed. Oxford: Pergamon Press.

Index

NOTE: *f.* indicates figure; *t.* indicates table.